D1001227

WITHDRAWN

782.1
G696a

Anna Bolena *and the Artistic Maturity of Gaetano Donizetti*

PHILIP GOSSETT

CLARENDON PRESS · OXFORD
1985

Oxford University Press, Walton Street, Oxford OX2 6DP

London Glasgow New York Toronto
Delhi Bombay Calcutta Madras Karachi
Kuala Lumpur Singapore Hong Kong Tokyo
Nairobi Dar es Salaam Cape Town
Melbourne Auckland

and associated companies in
Beirut Berlin Ibadan Mexico City Nicosia

Oxford is a trade mark of Oxford University Press

© Philip Gossett 1985

All rights reserved. No part of this publication may be reproduced,
stored in a retrieval system, or transmitted, in any form or by any means,
electronic, mechanical, photocopying, recording, or otherwise, without
the prior permission of Oxford University Press

British Library Cataloguing in Publication Data
Gossett, Philip
    Anna Bolena and the artistic maturity of Gaetano
    Donizetti.—(studies in musical genesis and structure)
    1. Donizetti, Gaetano. Anna Bolena
    I. Title   II. Series
    782.1'092'4   ML410.D7
    ISBN 0-19-313205-2

Library of Congress Cataloging in Publication Data
Gossett, Philip.
    Anna Bolena and the artistic maturity of
Gaetano Donizetti.
    (Studies in musical genesis and structure)
    1. Donizetti, Gaetano, 1797–1848. Anna Bolena.
2. Opera—Italy—19th century. I. Title. II. Series.
ML410.D7G67 1984   782.1'092'4   83–21993
ISBN 0-19-313205-2

Set by Eta Services (Typesetters) Ltd
Printed and bound in England by
Billing & Sons Ltd, Worcester

Contents

ALLEGHENY COLLEGE LIBRARY

85-6582

Editor's Preface

HISTORICAL musicology has recently witnessed vigorous and wide-ranging efforts to deepen understanding of the compositional procedures by which composers of various periods and traditions brought their works to realization. In part this trend has resulted from renewed and intensive study of the manuscript sources of works by many of the major figures in Western music history, especially those for whom new and authoritative editions of their complete works are being undertaken. In part it has arisen from the desire to establish more cogent and precise claims about the formative backgrounds of individual works than could be accomplished by more general stylistic study. In many cases, the fortunate survival of much of the composer's working materials—his sketches, drafts, composing scores, corrected copies, and the like—has stimulated this approach on a scale that could not have been dreamed of a century ago, when Gustav Nottebohm's pioneering studies of Beethoven's sketches and drafts first appeared.

Accordingly, this series aims to provide a number of short monographs, each dealing with an individual work by an important composer. The main focus will be on the genesis of the work from its known antecedent stages to its final realization, so far as these can be determined from the known sources; and in each case a view of the genesis of the work will be connected to an analytical overview of the finished composition. Every monograph will be written by a specialist, and, apart from the general theme, no artificial uniformity will be imposed. The individual character of both the work and the evidence, as well as the special viewpoint of the author, will inevitably dictate certain differences in emphasis and treatment. Thus some of these studies may stress the combination of sketch evidence and analysis, while others may place stronger emphasis on the position of the work within the conventions of its genre. Although no such series could possibly aim at being comprehensive, my hope is that it will deal with a representative number and variety of important works by composers of stature, including Bach, Mozart, Beethoven, Brahms, Wagner, Richard Strauss, Schoenberg, and Berg.

I take particular pleasure in inaugurating this series with this monograph by Philip Gossett on Donizetti's *Anna Bolena*. As a major American musicologist, as a well-known scholar of Italian opera of the nineteenth century, and as general editor of the current critical editions of the works of Rossini and of Verdi, Gossett needs no introduction. In this book he is able to demonstrate that Donizetti's *Anna Bolena* was much more than the work of a clever craftsman manipulating the received operatic traditions of his time. This study of the autograph manuscript of the opera shows that behind the stirring and brilliant surface of this work there lay conscious and painstaking revisions that were essential to its final form and that serve to reshape our understanding of the finished dramatic product.

Harvard University   LEWIS LOCKWOOD

Author's Preface

ENTERING the Archive of Casa Ricordi in Milan, those who love Italian opera must experience a sense of profound delight, mingled with awe. Everywhere the shelves overflow with original autograph manuscripts of works still dominating operatic stages throughout the world: a practically complete set for Verdi and Puccini, and a significant selection for Rossini, Bellini, and Donizetti. Accompanying these treasures is a vast selection of autographs and manuscript copies for operas by composers once famous but less well known or entirely forgotten now: the Mayrs, Pacinis, Mercadantes, Generalis, Pucittas, Riccis. Though their scores were enjoyed and praised by contemporaries, changing fashion and our own profound ignorance of works outside the current repertory have consigned them to oblivion. Some manuscripts represent composers obscure even in their own day, whose operas were performed without success, then abandoned. The entire history of an art form, its triumphs and failures, is revealed in this monumental archive, painstakingly collected and preserved by the publishing firm whose history is so intimately connected with the repertory.

It was at Casa Ricordi that I first studied intensively autograph manuscripts of nineteenth-century Italian operas. The firm remains a commercial music publisher, to be sure, and its employees are primarily concerned with supplying materials to performers, conductors, and opera houses. Yet their love of these documents, their fascination with the problems they pose, and their own skill at deciphering their meaning were crucial to the development of my awareness of what could be learned here. To Luciana Pestalozza, Fausto Broussard, and Luciano Petazzoni go my admiration and appreciation for years of assistance and friendship. I am no less grateful to the administrators of Casa Ricordi, especially Mimma Guastoni and Guido Rignano, for their constant support and kindness.

I prepared a preliminary study of Donizetti's *Anna Bolena*, whose autograph is found in the Ricordi Archive, for the *Primo convegno internazionale di studi donizettiani*, held in the composer's native town, Bergamo, during the summer of 1975, under the auspices of the Azienda Autonoma di Turismo and the Donizetti Society of England. It was an important moment for all who attended. The Donizetti revival had been raging since 1957, when the performances of *Anna Bolena* at the Teatro alla Scala in Milan, directed by Gianandrea Gavazzeni and starring Maria Callas, revealed that the operas of Donizetti could still excite modern audiences musically, dramatically, and emotionally. In the wake of this revelation, in the rediscovery of once-forgotten works, in the refutation of outmoded clichés copied endlessly from one book to another, a new generation of scholars and critics arose, fascinated by many aspects of Donizetti's art. The *Convegno* allowed them to meet, to exchange ideas, to hone their research techniques. My own work on Donizetti, and on Italian opera more generally, profited considerably from contacts begun or continued at Bergamo with Franco Arruga, William Ashbrook, Julian Budden, Franca Cella, Bruno Cagli, Rodolfo Celletti, Jeremy Commons, Francesco Degrada, Fedele d'Amico, Arrigo Gazzaniga, Friedrich Lippmann, Valeriano Sacchiero, and Alberto Zedda. I must add my gratitude to the late Herbert Weinstock for his indefatigable biographical studies and his personal friendship. This community of scholars concerned with Donizetti's music sustains us all.

The decision to expand my study of *Anna Bolena* into this monograph was prompted by Lewis Lockwood's agreement with Oxford University Press to oversee a series of studies of composer's autographs. Though I fear he was taken aback by the notion that I would write about Donizetti, when he had asked me to prepare a book on Beethoven's 'Pastoral' Symphony, I trust the resulting volume has helped demonstrate that even for a less exalted composer an autograph manuscript remains a fascinating document: if we know how to interpret its significance, it provides information and aesthetic insight impossible to achieve through other means. I am grateful to Professor Lockwood for a careful reading of this text, for encouragement and support that goes back to my student days at Princeton University, and for providing through his own work a model of what serious scholarship in music can be. Others who have read the manuscript at various times and provided useful suggestions include Ellen Harris, Jeffrey Kallberg, and H. Colin Slim.

During the final preparation of this study I had occasion to consult again important Donizetti manuscripts at the Pierpont Morgan Library in New York. If there is a more civilized library with a more helpful staff anywhere, I have yet to find it. My special thanks to Herbert Cahoon and J. Rigbie Turner for many kindnesses over the years and for making available photocopies of the important *Anna Bolena* autographs at the Morgan Library.

I first studied the autograph of *Anna Bolena* in 1971–2, with the assistance of a fellowship from the John Simon Guggenheim Foundation and a sabbatical leave from the University of Chicago. Let me take this opportunity to express my appreciation to both the Foundation and the University for their support.

My family has coped with much over the years, and to Suzanne, David, and Jeffrey go my love and thanks.

The University of Chicago   PHILIP GOSSETT

Introduction

DONIZETTI wrote quickly. In an active career spanning some twenty-five years, he produced more than sixty-five operas, not to mention numerous revisions, incomplete works, substitute arias, and non-operatic compositions. That comes to a yearly average of more than two-and-a-half operas, though during his most productive periods he regularly wrote four new works for the stage a year. Must we excoriate Donizetti for this fact? That depends on the point we are trying to make. When Hector Berlioz penned this paragraph on 16 February 1840, reviewing *La Fille du régiment*, which had its première at the Opéra-Comique of Paris on 11 February 1840, he was still suffering from the dismal failure of his *Benvenuto Cellini* at the Opéra a year and a half before:

What, two large scores at the Opéra, *Les Martyrs* and *Le Duc d'Albe*, two others at the Renaissance, *Lucie de Lammermoor* and *L'Ange de Nisida*, two at the Opéra-Comique, *La Fille du régiment* and another whose title is unknown, and still another for the Théâtre-Italien, all will have been written or transcribed in a year by the same author! M. Donizetti seems to treat us like a conquered nation, it is a veritable invasion. One can no longer say: the lyric theaters of Paris, but only: the lyric theaters of M. Donizetti. Never, in the days of his greatest favor, did the author of *Guillaume Tell*, *Tancrède*, and *Otello* dare show such ambition. He did not lack facility, to be sure, and he too had his baggage full [of earlier works that could be revised for Paris]. Yet, during his entire stay in France, he gave only four works and at the Théâtre de l'Opéra alone;[1]

[1] Berlioz's statement is not quite accurate. Rossini did produce four works for the Opéra, of which three were revisions of earlier operas: *Le Siège de Corinthe* (1826), a revision of *Maometto II* (1820); *Moïse* (1827), of *Mosè in Egitto* (1818); *Le Comte Ory* (1828), of *Il viaggio a Reims* (1825). Only *Guillaume Tell* (1829) was entirely new for the Opéra. At the same time, however, Rossini's operas made up the bulk of the repertory at the Théâtre-Italien, many produced under the composer's direction with new music expressly composed for Paris. They were featured at the Odéon in French adaptations, in at least one of which, *Ivanhoé*

Meyerbeer in ten years has mounted only two; Gluck, dying at an advanced age, willed to our principal theater only six large scores, fruit of a lifetime of work. It is true that they will endure for a long time.[2]

To the frustrated Berlioz, Donizetti's ready success in Paris was galling indeed. Yet when musicians discuss Mozart, the Austrian composer's extraordinary productivity, his ability to write the last three symphonies successively within three months becomes a sign of genius, of inexhaustible musical gifts. My purpose in drawing this comparison is not to pretend that Donizetti's operas are on the same aesthetic plane as Mozart's last symphonies. It is rather to state unequivocally that the aesthetic quality of a work of art, the extent to which it has endured or will endure, is not a function of its period of gestation. Let us purge this diction from our critical vocabulary.

Though Donizetti wrote quickly, he did not write carelessly. He was a thorough professional, who knew his *métier* and lived at a time in which composers of Italian opera depended upon a coherent set of structural conventions against which the particularities of their style could emerge. It was not a matter of merely plugging into a ready form and churning out pre-composed, pre-digested musical structures, with little or no concern for their individuality. Analysis of completed works teaches us much, but only analysis of composers' autographs and sketches brings us into direct contact with the composer as a creator. From such study we know that Donizetti's manuscripts were often a battleground. These canceled measures, substituted folios, altered vocal lines, tormented sketches: do they reveal a composer who worked mechanically or with unthinking facility? If Donizetti can be caricatured, as he was in France during the 1830s, writing one opera with his right hand and another with his

---

(1826), Rossini had some hand. And he composed for the coronation of Charles X in 1825 a complete Italian opera, first performed at the Théâtre-Italien, *Il viaggio a Reims*, later adapted for the Opéra to a new libretto. Certainly a critic in the mid-1820s could have complained as bitterly about Rossini's hold on French musical life as Berlioz complained about Donizetti's in 1840.

[2] The article is reprinted in Hector Berlioz, *Les Musiciens et la musique* (Paris, n.d.), pp. 145–56; see particularly p. 152. All translations throughout this study are my own unless otherwise specified.

left,[3] we had best provide him with another pair of hands to make corrections in the scores as he proceeded.

Why, after all, do we study a composer's autograph manuscripts? There are many possible answers, the least controversial of which is: we use them to produce accurate musical editions, as close to the composer's intentions as possible. Yet anyone who has worked with autographs knows their limitations in this sphere. Most eighteenth- and nineteenth-century autographs require significant 'interpretation' when used as the basis for an edition. Those of Donizetti are no exception. Rarely do they show precisely how the music should be performed, let alone edited: orchestration of music for the stage *banda* is usually absent; some instrumental parts, such as the bassoons, are incompletely notated; vocal lines lack the appropriate ornamentation necessary to bring them to life; articulation is coherently presented in neither vocal lines nor orchestral parts. Though no other sources may be as close to Donizetti's wishes as his autograph manuscripts, their interpretation requires extensive knowledge of the context in which they were written.

We also examine a composer's manuscripts to learn something about his biography and the 'biography' of the work of art. Through recent advances in the study of manuscripts, it has been possible to recast the artistic development of J. S. Bach, to gain control over the chronology of many early pieces by Beethoven, to place in the context of Mozart's career his surviving fragmentary manuscripts. We know when Schubert composed most of his major works, and can even specify that certain sections of music were added to pre-existent compositions years later. By studying the autograph score of Rossini's *L'Italiana in Algeri* we have determined how the work was composed, what sources were used for its libretto, which sections were added by other composers and why, and which ones Rossini later revised and

---

[3] See the magnificent caricature published in the *Panthéon charivarique*, reproduced in Guido Zavadini, *Donizetti. vita-musiche-epistolario* (Bergamo, 1948), facing p. 512. It is worth noting the quatrain that appears under the caricature:

> Donizetti dont le brillant genie
> Nous a donné cent chef d'oeuvres divers
> N'aura bientot qu'une patrie
> Et ce sera tout l'univers.

when. That such information is crucial to the 'biography' of Rossini and of his opera seems clear.[4]

Most controversial remains the position that we study autograph manuscripts better to comprehend the aesthetic qualities of individual pieces. Some writers hold that to understand a work of art we must analyze it; for this purpose we need consult only the work itself. By 'analysis' they mean primarily linear analysis of the kind practiced by Heinrich Schenker and his disciples. Enough ink has been spilled on this subject over the past few years, particularly in the realm of Beethoven studies, that it seems useless to pursue the battle further here.[5] Italian opera has largely escaped from the controversy because it is generally deemed unworthy of being 'analyzed'. Perhaps that is a blessing. Instead of coming to Donizetti's music with a massive theory that defines what is important and how it should be studied, we can attempt to generate appropriate methods for this particular repertory. What kind of an opera is *Anna Bolena*, why is it written the way it is, and how can we best appreciate its unique qualities? There is no more effective way to approach these questions than by investigating Donizetti's manuscript. In it the composer defines for himself and for us a wide range of aesthetic problems. He compares versions, makes choices, accepts one procedure, with its inherent aesthetic significance, rather than another. Donizetti does not explain his choices. We must evaluate them for ourselves and derive from them criteria by which the composer may have been guided. We must then apply these criteria to other situations, measuring them against other works by Donizetti and his contemporaries. Though we cannot hope for absolute 'truths' in such a venture, Donizetti's autograph poses a challenge we cannot lightly dismiss: to engage the composer in direct dialogue. Can we enter, even

---

[4] There is no point in citing here the extensive literature on Bach, Beethoven, Mozart, and Schubert of the past few decades. Analysis of the 'biography' of *L'Italiana in Algeri* is found in the introduction to the critical edition of the score, edited by Azio Corghi as part of the *Opera omnia di Gioachino Rossini*, published by the Fondazione Rossini (Pesaro, 1981). See particularly pp. xxi–xxxi.

[5] The most extreme position is that of Douglas Johnson, stated in his 'Beethoven Scholars and Beethoven's Sketches', *19th-Century Music*, ii (1978), 3–17. Two very different responses are given by Joseph Kerman, 'How We Got into Analysis, and How to Get Out', *Critical Inquiry*, vii (1980–1), 311–31, and Philip Gossett, 'The Arias of Marzelline: Beethoven as a Composer of Opera', in *Beethoven-Jahrbuch*, x (1983), 141–83.

hypothetically, into his mode of thought, recognize his motivations, understand the process of composition sufficiently to penetrate the criteria, the goals, the aesthetic world of the composer himself?

Let me emphasize that I am not advocating a brand of extreme historicism, admitting as appropriate for study of a work of art only those approaches likely to have been familiar to its composer. We have all learned too much from modern analytical methods to engage in such a pointless crusade. Musical compositions can be approached from a multiplicity of viewpoints, and our understanding depends on the breadth, variety, and detail with which we explore them. Because of the particular intimacy it creates between the composer and the critic, however, I find the approach through analysis of a composer's autograph manuscript unusually fruitful. That approach will be basic to this book. It does not pretend to exhaust the analytic or aesthetic issues surrounding this or any work of art, for it is limited by its very methodology to those matters for which we possess written evidence. The question is not whether our approach provides a complete understanding of the work, an absurdity in any event, but only whether it offers insights that contribute significantly to how we conceive and hear a musical composition. That some aspects, not evident in the written record of the creative process, will be inadequately covered should lead our research into other, parallel areas of investigation.

It is a critical commonplace that *Anna Bolena* marks a crucial turning point in the life and art of Donizetti. This study will trace some of the historical roots of this evaluation and examine the foundation on which it rests. We will assess the kinds of compositional questions with which Donizetti was wrestling in 1830 and the nature of the decisions he made. The implications of these questions and decisions are fundamental for a proper understanding both of *Anna Bolena* and of Donizetti's attempt to forge a personal style, a style independent of the operatic conventions that dominated contemporary Italian opera, those associated with the works of Gioachino Rossini.

I. The History

GAETANO DONIZETTI was born in the northern Italian city of Bergamo on 29 November 1797. Although his earliest full-length opera was written in 1818, his compositional career did not begin to flourish until 1822, with the extremely successful première of his *opera seria, Zoraida di Granata*, at the Teatro Argentina of Rome on 28 January. Between this work and *Anna Bolena*, whose première took place almost exactly nine years later, Donizetti composed twenty-three more operas, all but one of which were immediately performed: two for Rome, one each for Palermo, Genoa, and Milan, and fully seventeen for the theaters of Naples. His early successes with the Neapolitan audience were so noteworthy that in 1827 he signed a formal contract with the impresario, Domenico Barbaja, to compose four operas each year for a period of three years and to assist in the artistic administration of the Neapolitan theaters, a position similar to one held by Rossini a decade earlier, though the Pesarese generally composed only a single opera each year for Barbaja.

The reception accorded Donizetti's operas in northern Italy was much less enthusiastic. His one work for Milan, *Chiara e Serafina*, first performed at the Teatro alla Scala on 26 October 1822, was a resounding failure. As late as 1830 he was keenly sensitive to this situation. Donizetti's teacher was the prolific Bavarian composer of opera and church music, Giovanni Simone Mayr, most of whose career was spent in Italy, where he ultimately settled in Bergamo and became *maestro di cappella* at the cathedral of S. Maria Maggiore and director of the city's music school. During the Carnival season of 1830, Mayr prepared a revival of his prize pupil's most successful comic opera to that time, *L'ajo nell'imbarazzo*, first performed in Rome at the Teatro Valle on 4 February 1824. It met with scant favor despite Mayr's pains. News of the failure was communicated to Donizetti by his father, to whom the composer replied from Naples on 13 February 1830:

Bravo, Bravo, Bravo! I truly take a crazy delight that *L'ajo* was such a fiasco; I am sorry only for the pains of the good, or rather wonderful, Mayr; for the rest, I laugh at it. Up there, everyone whistles, while here I receive applause . . .

For my benefit evening I wrote a Comedy in the form of a *farsa*, *I pazzi per progetto*, which succeeded brilliantly, perhaps because I am looked upon with favor, but everything I do here goes well. You people in Lombardy should simply never perform my things, never, never; the newspapers have discredited me too much, and you may as well let sleeping dogs lie. [Z. No. 53][1]

His wounded pride is everywhere apparent.

Donizetti's plans for the remainder of the year included two new operas for Naples, one imminent, *Il diluvio universale*. This three-act *azione tragico-sacra* on the story of Noah and the Ark, with a sub-plot of love, duty, and conversion not unlike the one in Rossini's similar *azione tragico-sacra* a decade earlier, *Mosè in Egitto*, was performed on 28 February 1830 at the Teatro San Carlo. It was followed late that summer by *Imelda de' Lambertazzi*, an *opera seria* in two acts, premièred at the same theater on 5 September. From a letter to his father dated 4 May [Z. No. 55], we know both that Donizetti was beginning *Imelda* and that he was in negotiations with Rome for a new opera to be given during the Carnival season of 1831. On 29 May he reported further to his father: 'For Carnival I may well write an opera in Rome, and therefore I have begun working early here so I can be free sooner: I'll present *Imelda de' Lambertazzi* here at S. Carlo' [Z. No. 56]. By 24 June his plans had changed, as he told Mayr: 'I was supposed to write for Carnival in Rome, but I prefer not to write a Buffo opera because the company there is weak. Nothing else to report' [Z. No. 58]. Soon, however, there was a great deal to report.

Direction of the modest Teatro Carcano of Milan had been assumed by a group of noblemen and dilettantes who were determined to mount a truly extraordinary season of opera in competition with the offerings at the Teatro alla Scala. Towards

[1] Unless otherwise stated, all letters cited are taken from the Zavadini collection, op. cit., and from the supplement, edited by Guglielmo Barblan and Frank Walker, published as *Studi Donizettiani*, i (1962). They will always be identified by their number in that collection, their 'Z' number.

this end, they assembled a stellar group of singers, including Giuditta Pasta and Giovanni Battista Rubini, and contracted the two leading young composers in Italy, Donizetti and Bellini, to prepare new operas expressly for performance at the Carcano. The negotiations with Donizetti were largely completed by 20 July, for on that day he responded to yet another request for his services during the forthcoming Carnival season, this one from the Teatro La Fenice in Venice:

I have a contract with the Impresarios of the Carcano to compose the first opera for the coming Carnival, and after this (should they grant me the payment requested) I must also remount another opera of mine, either *buffa* or *seria*, at their discretion, whose production, however, must occur no later than the month of January 1831. [Z. No. 59]

Although Donizetti attempted to devise an agreement that would have permitted him to work for Venice after the Carcano season, this proved impossible. He actually signed the contract with the Carcano on 1 August, as he informed Mayr on 7 August:

Finally just six days ago I signed the contract with the Carcano, for 650 *collonati*, with trip and lodging paid, that's not bad. . . . After my *Imelda* is in place, I will fly to the home of risotto [Milan] and then to that of *polenta* and game birds [Bergamo], and what a feast we must make, we really must. [Z. No. 60]

Imelda had its première on 5 September; soon after, Donizetti and his young wife Virginia left for Rome, where she was to remain with her family for the next few months. By 5 October Donizetti was in Bologna, from where he communicated to his father:

On Sunday [10 October] I will be in Bergamo, and I beg you not to trouble yourself to come to Milan, for fear we might miss one another on the road. Tomorrow I leave here, and in Milan I will look for my lodgings, deposit my things, then with the express coach I will come post haste. I don't know at what hour I will arrive, but I think early.
    Don't be alarmed if I am delayed, for who knows whether or not I will have to spend some time with the poet in Milan, but I hope not. [Z. No. 61]

He did go to Bergamo, but we do not know when nor for how long he stayed.

    At this point documentary sources concerning the preparation of *Anna Bolena* fail us. All Donizetti biographies agree, though they cite no evidence, that Felice Romani finished the libretto by 10 November and that until 10 December Donizetti was a guest of the *prima donna* Giuditta Pasta at her villa on Lake Como. According to these reports, he returned to Milan with a completed manuscript to start rehearsals that day.[2] The cast included Pasta as Anna and Rubini as Percy. Filippo Galli, who had created so many roles for Rossini, including Selim in *Il turco in Italia* (1814), the title role of *Maometto II* (1820), and Assur in *Semiramide* (1823), was Enrico, and Elisa Orlandi was Giovanna Seymour. The reviewer in the *Gazzetta di Milano* of 27 December gave this opinion of the première, which took place the previous day, the traditional opening of the Carnival season:

The cabaletta of an aria sung by Rubini . . . the animated idea in the finale: here is the essence of what truly pleased in the music of the first act. The other applause that was heard was due only to the singers.

But in the second act the affair changed aspect, and the talent of the maestro demonstrated itself with uncommon strength. A duet, a trio, and three arias are of beautiful and grandiose structure. As for the performers, one has to have heard Pasta and Rubini in their two arias of differing type and pattern to form an idea to what point the power of declaimed song and the enchantment of perfect sounds may attain. In the trio Pasta, Rubini, and Galli showed themselves not only at their best, but truly masters in the perfection of their ensemble. This is all the more notable to the extent that in this trio, rendered very complex by its beautiful harmonies, Donizetti has revealed himself a worthy and favorite pupil of Mayr.[3]

Several days later Donizetti assisted at the staging of his *opera semiseria*, *Gianni di Calais*, first performed in Naples at the Teatro del Fondo on 2 August 1828, an opera originally composed for the same Rubini.

---

[2] See the reports given in the major Donizetti biographies. These include: William Ashbrook, *Donizetti and his Operas* (Cambridge, 1982), p. 63; Herbert Weinstock, *Donizetti and the World of Opera in Italy, Paris and Vienna in the First Half of the Nineteenth Century* (New York, 1963), p. 72; and Zavadini, op. cit., pp. 38–9. It would be fascinating to know more about Felice Romani's own working methods or about any direct contact between him and Donizetti. Unfortunately no one has thus far discovered the location of Romani's papers, if they have survived at all. That such materials would be of enormous interest is clear from the papers of Salvatore Cammarano, preserved in the Biblioteca Lucchesi-Palli of Naples, papers that document many contacts between Donizetti and his Neapolitan librettist.

[3] Cited after Ashbrook, op. cit., p. 64, in his translation.

Donizetti remained in Milan throughout the rest of January 1831. During this period he made significant alterations in *Anna Bolena*. These were alluded to in the following notice from *L'eco*, another Milanese paper, of 4 February: 'This opera has been corrected in some places by M.° Donizetti. In the first performances (after an interruption) we will hear changes in it which we hope will bring greater glory to the composer'.[4] Donizetti was no longer in Milan, however, having departed on 31 January for Rome, where he rejoined his wife and her family. From there, they continued to Naples, concluding the trip that established the composer's reputation in northern Italy.

[4] Ibid.

II. Critical Evaluations of Anna Bolena

Anna Bolena quickly spread its composer's fame even further. Although not the first of Donizetti's operas to be performed outside Italy, it was the first to achieve international acclaim. Donizetti's earlier reputation in France, where Italian opera flourished during the 1820s, was succinctly put by François-Joseph Fétis, who began publication of his *Revue musicale* in 1827 with a survey of contemporary music. Fétis divided Rossini's followers into various grades of quality: the best were Carafa, Mercadante, and Meyerbeer; the next group included Pacini and Vaccaj; finally, 'In the lowest position among living composers are found Donizetti, Raimondi, Sapienza, obscure imitators of the Rossinian manner. It is useless to list their works; they rarely emerge from the cities for which they were written'.[1] Donizetti had already composed about twenty operas.

When *Anna Bolena* was first performed in Milan on 26 December 1830, the notice appearing in the *Revue musicale* was mixed:

The first act seemed cold, with the exception of the Finale, which is reported to be well constructed; in the second act there is more warmth. A duet between Anne Boleyn and Jane Seymour, the stretto of the Trio, 'Salirà d'Inghilterra sul trono', the Aria of Percy, and the final scene garnered some applause.[2]

---

[1] François-Joseph Fétis, 'Examen de l'état actuel de la musique en Italie, en Allemagne, en Angleterre et en France,' *Revue musicale*, i (1827), 88.

[2] *Revue musicale*, x (1830–1), 310, issue of 15 January 1831, as part of the 'Nouvelles Étrangères.' Fétis was simply reporting what he had gathered from a none-too-reliable Milanese source, who, for instance, assigned 'Vivi tu', Percy's famous aria, to Giuditta Pasta. I am calling attention here to several reviews of *Anna Bolena* that have gone unmentioned in the secondary literature. For a précis of Milanese criticism of the time, see Ashbrook, op. cit., pp. 63–5. Ashbrook also quotes from the English critic, Henry Chorley (see p. 66).

Fétis's own response to the opera, after he heard its Parisian première later that year, was more interesting, as we shall see. In Germany the Milanese correspondent of the *Allgemeine musikalische Zeitung* sent a dispatch before actually hearing the opera. It appeared in the issue of 16 March 1831, affirming that 'Much in the second act, especially a Trio, has been praised'.³ A full review followed only on 18 May:

(Teatro Carcano). Donizetti's *Anna Bolena*, about which we already published a report based on hearsay, has been so successful that it is unanimously acclaimed the most beautiful opera given in either theater [the Carcano or the Scala] during this season, playing always before a full house. It is well known that Donizetti belongs among today's great favorites, and in the first act of *Anna Bolena* he does little to distinguish himself from them; but in the second act he emerges as himself, and shows himself to be a worthy student of Mayr. It is to be hoped that he extensively revises the first act, never departing again from his new and individual path. Unfortunately there is a large fly in the soup; probably it was only in obedience to convention that he allowed himself to conclude the beautiful Trio with an ordinary cabaletta.⁴

There are several aspects of this review to which we shall return.

Anna Bolena was the first Donizetti opera to be performed in Paris. The cast of the revival, which opened on 1 September 1831, included Pasta and Rubini, both members of the original cast, with Luigi Lablache replacing Galli in the role of Enrico and Eugenia Tadolini replacing Orlandini as Giovanna Seymour. Fétis's review, though it reveals his ignorance about many details of the composer's life (he claimed, for instance, that Donizetti was born in Rome), reflects the critical stance that was from then on to be taken by many other writers towards the opera. Though he probably had heard none of Donizetti's earlier efforts, Fétis none the less felt free to pronounce judgement on them. Donizetti, a student of Mayr's, he began, is one of the best young Italian composers, 'perhaps he whose style is the purest', but

... even with these advantages and with ideas unique to his own sensibilities, during most of his life he has produced works that are only

---

³ *Allgemeine musikalische Zeitung*, xxxiii (1831), column 171.

⁴ *Allgemeine musikalische Zeitung*, xxxiii (1831), columns 323–4. In this same season the Teatro alla Scala opened with Bellini's *I Capuleti ed i Montecchi*, in its first Milanese performance, while the Teatro Carcano also presented the première of his *La sonnambula*.

more or less mediocre. Seduced by Rossini's successes and imagining no better way to equal them than by imitation, he has not troubled to search within himself for those qualities granted him by nature; rather, with that insouciance most second-rate Italian composers show towards their works, he has preferred to consult his memory rather than his genius. Written with neglect and a prodigious speed, his early works followed one another without doing a thing for his glory. Scarcely had he finished in one city before he ran to another, often repeating in the latter what he had written for the first. Fortunately, he himself had little esteem for works prepared in this way. Though a contract with Barbaja to become director of music and composer for the Neapolitan theaters obliged him to multiply his new works excessively, the need to sustain himself before an exigent public caused him at least to write more carefully and to seek inspiration from within himself more than he had heretofore done. From this moment date his finest works, and particularly *Anna Bolena*.

It is not that this opera is irreproachable under the heading of imitation; there is much Rossinism there still, and Donizetti, by the uniformity of his means, the similarities one notices among many phrases, and especially the immoderate use of noisy instruments, has fallen into one of the worst faults that can stain art works, that is, monotony. Still, it is undeniable that there are beautiful things in this score. The Aria of Percy (with *pertichini* [interventions from other, secondary, characters]), the Duet of Jane Seymour and Henry VIII, part of the first-act Finale, the scene of Anne Boleyn and Jane Seymour, a period of the Trio in the second act, numerous fragments from the second-act Finale, and several other phrases scattered here and there in the rest of the work, all announce an imagination rich enough to produce a fine opera, if he who possesses this quality has also the unwavering will to employ it.[5]

As I suggested above, it is patently absurd that critics, old and new, blame the failure of Donizetti's early operas (not to mention the operas of Pacini or even lesser contemporaries) on the rapidity of their creation and the attendant theatrical circumstances, yet marvel that *Il barbiere di Siviglia* was composed in two weeks. There is absolutely no evidence that Donizetti spent a day more on the creation of *Anna Bolena* than on *Imelda de'Lambertazzi* earlier that year. 1833 saw the premières of *Il furioso all'isola di San Domingo, Parisina, Torquato Tasso,* and *Lucrezia Borgia*. In no earlier year had Donizetti written more new music. Whatever

<hr>

[5] *Revue musicale*, xi (1831), 249, issue of 10 September 1831.

differentiates Donizetti's operas after *Anna Bolena* from his earlier ones, it is certainly not how long it took him to compose them. Fétis's view of Donizetti at the time of *Anna Bolena* is remarkably close to the propagandistic vision set forth by Giuseppe Mazzini in his *Filosofia della musica* (1836). After discussing both *Anna Bolena* and *Marino Faliero*, Mazzini cited aspects of these operas which are

... potent indications of a talent who has not yet fully developed, who catches a glimpse of a new musical world, who would like to pursue it, who, because he is entangled, strangled by a thousand obstacles that today block a capable talent, does not pursue it; but who has none the less revealed himself in first steps, on the basis of which future generations will, I believe, be able to say: *He would have been powerful enough to conquer the new world, had he truly wished to do so.*[6]

Mazzini retraced Fétis's arguments, describing the earlier works as saturated with Rossinian devices. Though his view of *Anna Bolena* is well known, it deserves to be quoted in full again:

The individuality of the characters, so barbarously neglected by the servile imitators of Rossini's lyricism, is painted with rare energy and religiously observed in many of Donizetti's works. Who has not felt in the musical expression of Henry VIII the language, severe, tyrannical, and artificial at the same time, that history assigns to him? And when Lablache fulminates these words:

> Salirà d'Inghilterra sul trono
> Altra donna più degna d'affetto, etc.

who has not felt his spirit shrink, who has not understood all of tyranny in that moment, who has not seen all the trickery of that Court, which has condemned Anne Boleyn to die? And Anne, furthermore, is the resigned victim, whom the libretto—and history too, whatever others may say—depicts, and her song is a swan-song that foresees death, the song of a weary person touched by a sweet memory of love.

This *Anna Bolena* approaches epic poetry in music. Smeton's *romanza*, the duet of the two rivals, Percy's 'Vivi tu', etc., Anna's divine 'Al dolce guidami', and, in general, the concerted pieces, irrevocably place this opera among the first in the repertory. The orchestration, if not yet

---

[6] Giuseppe Mazzini, *Filosofia della musica* (1836), edited with an introduction by Adriano Lualdi (Fratelli Bocca, 2nd ed., 1954), p. 185. A useful introduction to this book is provided by Albert Seay in his article, 'Giuseppe Mazzini's *Filosofia della musica*', *Notes*, xxx (1973–4), 24–36.

equal to the melodic inspiration, none the less is full, continuous, majestically solemn. The choruses, among which must be noted especially the 'Dove mai n'andarono', etc., give the work a finish that, within the limits of our present capabilities, leaves nothing to be desired.[7]

The prophetic force of Mazzini's prose was not lost on those who saw the career of Giuseppe Verdi as the natural fulfillment of Donizetti's legacy.

Half a century later, Emilia Branca attributed a lengthy statement to her late husband, Felice Romani, librettist of *Anna Bolena*. When artists of the quality of those who sang its première are available,

. . . then the poet, throwing aside the faded, melodramatic jokes called *libretti*, achieves the heights of lyrical tragedy; then a composer, leaving in his trunk his normal repertory of old motives and endless cabalettas, raises himself to dramatic truth and music of passion: then finally appears an ANNA BOLENA.[8]

One wonders what opera Romani is talking about, for however much we may applaud the superior qualities of *Anna Bolena*, every one of its musical units concludes with a cabaletta.

Modern commentators have generally adopted similar positions. Zavadini, discussing *Imelda de' Lambertazzi*, wrote:

We must none the less agree that this opera, like all its predecessors, is overly dependent on the influence of Rossini and therefore, whether appropriately or not, like them was soon forgotten. From this moment on, however, the author sought to establish his own personality and to our grand fortune he finally discovered it in his very next opera.[9]

Barblan presented a view of the work much in the style of the original reviews, commenting, for example, about the second-act Finale: 'The most intense Romantic shudder, in its funereal aspect, beat at the doors of Donizetti's heart: the Rossinian outlines dissolved and disappeared in the tempest of his genius'.[10] In an article about the opera, Ashbrook reported: 'The opening

---

[7] Mazzini, op. cit., pp. 179–81.

[8] Emilia Branca, *Felice Romani ed i più riputati maestri di musica del suo tempo* (Ermanno Loescher, 1882), p. 230.

[9] Zavadini, op. cit., p. 38.

[10] Guglielmo Barblan, *L'opera di Donizetti nell'età romantica* (Bergamo, 1948), p. 54.

night was an historic triumph, Donizetti conclusively demonstrating that he had finally worked his way clear of Rossinian idiom and found a personal style'.[11]

Critical opinion of *Anna Bolena*, it would seem, changed little from 26 December 1830 until now. Only in Ashbrook's recent revision of his 1965 *Donizetti* does a more balanced picture begin to emerge:

> Increasing familiarity with the operas Donizetti wrote before *Anna Bolena* reveals that achievement in a new perspective as the logical culmination of tendencies present in his work from the beginning, and not as a sudden, scarcely anticipated step from mediocrity into greatness, not as the sudden throwing-off of a consciously adopted *rossinianismo* from which he emerged abruptly as his own man.[12]

Yet even here the precise elements that distinguish the opera from its predecessors or that it shares with them remain obscure.

★

The vision of *Anna Bolena* set forth in these statements is, to be sure, an appealing one. The opera that first placed Donizetti's star in the European firmament emerges as a work in which the composer's mature style is first clearly manifest. That practically none of the contemporary critics knew Donizetti's earlier music is irrelevant. Nor do we ever need to learn what actually make up those Rossinian conventions from which Donizetti is said to be escaping. They are simply flung up, a massive, towering system that would account for everything in Rossini from *Demetrio e*

---

[11] William Ashbrook, 'Anna Bolena', *The Musical Times*, cvi (1965), 433.

[12] Ashbrook, *Donizetti* (1982), p. 47. One aspect of Ashbrook's discussion that needs further development is his assertion that many melodic ideas in *Anna Bolena* were borrowed from earlier operas, including *Enrico di,Borgogna* (1818), *Gabriella di Vergy* (1826), *Otto mesi in due ore* (1827), *Il paria* (1828), and *Il diluvio universale* (1830). (See, *passim*, pp. 54, 60, 216, 231, and 300.) 'Borrowing' is an elusive term, and until we know just what is involved in each of these cases it is difficult to gauge their importance. That the first four measures of a lyrical section in *Enrico di Borgogna* anticipate the tune of 'Al dolce guidami' is little more than a curiosity, if we are to believe Ashbrook that the resemblance goes no further. That a melody in *Gabriella di Vergy* was used 'as the basis of' a piece in *Anna Bolena* can mean everything or nothing (see p. 300). An urgent task of Donizetti research in general should be careful study of the problem of 'self-borrowing'.

Polibio to *Guillaume Tell*. Whether such a view is justified seems to be equally irrelevant. The brand of criticism normally applied to *Anna Bolena*, as to most of Donizetti's music, has never depended on either a firm knowledge of the operas or a thorough investigation of the composer's style. It is generic, rather than specific criticism: criticism depending for its effect on sweeping visions, unwilling to linger over detail in its pursuit of larger truths, and, ultimately, impervious to anything but its own rhetorical needs.[13]

The resulting portrait of Rossini is of particular interest. He is transformed from the outstanding musical figure of the early nineteenth century in Italy, the Napoleon of music, into a villain, a child molester, responsible for all that is evil in later Italian opera, ravishing those unlucky enough to be born after him. Rarely does critical writing about Donizetti permit us to distinguish one of this composer's early operas from another or to comprehend just how any of them are really related to any particular Rossini operas. Bedevilled by this criticism, further-more, one could hardly help but believe, to take a single example, that Rossini never differentiated characters in a duet. Now it is perfectly true that the archetype of the Rossinian duet does not encourage character specificity, but one need not search far to find prominent examples in which Rossini adapts his archetypes to this end. The opening section of the second-act duet in *Otello* between Jago and Otello, 'Non m'inganno', is an obvious instance. Though Rossini began with two poetically parallel stanzas, which could conceivably have been set to the same music, he chose another, more dramatically effective technique: Otello reads aloud the letter he presumes Desdemona sent to Rodrigo, while Jago provides a running commentary, line by line. Even in the cabaletta of a duet, Rossini at his best can portray the emotions of individual characters. In 'Tu serena intanto il ciglio', which

---

[13] Among the few significant studies that go further should be singled out the volumes by Ashbrook and Barblan, as well as the articles of Rodolfo Celletti, 'Il vocalismo italiano da Rossini a Donizetti,' *Analecta Musicologica*, v (1968), 267–94; vii (1969), 214–47, and Friedrich Lippmann, 'Die Melodien Donizettis', *Analecta Musicologica*, iii (1966), 80–113. Piero Rattalino's 'Il processo com-positivo nel "Don Pasquale" di Donizetti', *Nuova rivista musicale italiana*, iv (1970), 51–68 and 263–80, shows very little understanding of the fascinating problems in that autograph.

concludes the *Semiramide* duet, 'Ebben, a te, ferisci!', instead of each character singing through the tune once, the archetypical design, the melody is divided between them. Arsace's more harmonically conceived opening phrase, attempting to calm his mother's fears, contrasts sharply with Semiramide's hysterical, more chromatic response, in which the reiterated half-step from C♯ to D significantly raises the level of tension:

Ex. 1. Beginning of the cabaletta theme from the Duetto for Arsace and Semiramide, 'Ebben, a te, ferisci!' from *Semiramide*

In *Guillaume Tell*, of course, there are many examples of this kind: the duet for Guillaume and Arnold in Act I, the Act II trio with Arnold's heart-rending 'Ses jours qu'ils ont osé proscrire', and so forth.[14]

Rossini was unquestionably responsible for codifying certain structural, melodic, and rhythmic devices that played an enormous part in defining Italian opera throughout the first half of the nineteenth century. No musician in Europe at the time was ignorant of this phenomenon, and composers were keenly aware that they were working under the shadow of a colossus. Giovanni Pacini, an outstanding Italian composer of the period, described a certain group of his own works in the following terms:

[14] I have tried to suggest a model for defining archetypes in this music and employing them in both critical writing and studies of authenticity in my 'Le sinfonie di Rossini', *Bollettino del centro rossiniano di studi* (1979), 5–123. An excerpt from this monograph is published as 'The overtures of Rossini' in *19th-Century Music*, iii (1979), 3–31.

I will say, therefore, without falsifying the truth, that I always followed the same system, though I abandoned crescendos and always sought new forms and structures for numbers, however difficult these were to find. But may I be permitted to observe that my contemporaries all followed the same school, the same procedures, and consequently were imitators, just as I was, of the *Brightest Star*.—But, good Heavens! what else were we to do to sustain ourselves? If I was therefore a follower of the great Pesarese, so were the others. They might have been more fortunate in their melodic ideas, more accurate in their instrumentation, more learned; but the form and structure of their pieces were similar to those of mine.[15]

Yet it is precisely the function of serious criticism to define what particularizes each composer, each opera, and each number. To the extent we succeed, we begin to understand a composer's style; to the extent we understand, we can place an individual work, such as *Anna Bolena*, into a more accurate historical framework.

Another point needs to be made about the Rossinian conventions. Critics use 'conventional' as a term of disapproval; rarely does it have its proper sense, that is, exemplifying a convention. Though the reviews cited above are unanimous in finding parts of *Anna Bolena* to be derived from Rossinian models, different critics find different pieces 'conventional'. The *Revue musicale* and Mazzini refer in laudatory terms to 'Salirà d'Inghilterra sul trono', the cabaletta of the trio, 'Ambo morrete, o perfidi'.[16] The *Allgemeine musikalische Zeitung*, on the other hand, calls this cabaletta the 'fly in the soup', a mere bow to 'convention'. They are, of course, both correct: it does exemplify a convention, but it is also a magnificent and particularly appropriate use of that convention, in both musical and dramatic terms. Need it be reiterated that there is no more formally conventional number in the entire Donizetti canon than the Finale to Act II of *Lucia di Lammermoor* (which includes the improperly named 'sextet')? The piece is a text-book example of the Rossinian 'Finale', even if it happens to be the best exemplification of that archetype in the post-Rossini period. Using the term 'conventional' as one of opprobrium is simply unacceptable: much of Donizetti's finest

---

[15] Giovanni Pacini, *Le mie memorie artistiche* (Firenze, 1875), ed. by Ferdinando Magnani, p. 54.

[16] There are several textual problems in this cabaletta, to be discussed below, but they are unlikely to be at the heart of this critical dispute.

music (not to mention Verdi's or, for that matter, Mozart's) is worked out within the most rigidly conventional boundaries.[17] A final problem must be raised before we return to *Anna Bolena*. It is still practically impossible to 'know' the operas Donizetti wrote in the 1820s (not to speak of later ones), because our control over their musical texts is appallingly inadequate. Assembling and ordering this material will be an arduous task, but it is an essential one for students of Italian opera. That our ignorance extends to major works can be seen in the case of *Anna Bolena*. Only failure to consult Donizetti's autographs can explain the absence of reference in the Donizetti literature to the existence (in the autograph score) of a duet for Anna and Percy that replaces 'S'ei t'abborre', the duet opening the first-act Finale. Donizetti probably wrote this magnificent new piece, 'Sì, son io che a te ritorno', before the end of the first Milanese season.[18]

I have tried thus far to cast doubt on the accepted critical view of *Anna Bolena*. I do so not because I doubt the essential truth of that view, but because in its extreme subjectivity it denigrates works that are largely unknown, affirming historical patterns and relationships without solid foundation. Establishing such a foundation will be an arduous task, requiring the co-operative efforts of many scholars. We lack even a generally accepted language to describe the most basic musical procedures in Italian opera of this period. Little wonder that comparisons between one work and another or one piece and another, even those that do not limit themselves to rhapsodic prose, tend to lack enough precision to be truly helpful. Yet it can be demonstrated that in certain respects *Anna Bolena* is indeed a work in which Donizetti attempts to define a personal style. These demonstrable elements do not account by any means for all aspects of the opera, but they play a significant part in giving the opera its particular character. As such, they demand our attention.

[17] Ashbrook, *Donizetti* (1982), states this well: 'The merit of *Anna Bolena* lies not in any radical departures from Donizetti's previous norms of structure, but rather in the generally higher level of expression to which his imagination was stimulated by the well-realized characters he was bringing to life.' (p. 317)

[18] The new Duetto, frequently performed in place of the original piece over the next few years, will be the subject of detailed scrutiny in Chapter XI. When the length of the opera prompted singers to omit a duet here altogether, the composer reused part of 'Sì, son io che a te ritorno' in *Marino Faliero*, where it became the Duetto for Elena and Fernando, 'Tu non sai la nave è presta'.

ALLEGHENY COLLEGE LIBRARY

III. Donizetti as Reformer (1830)

FORGING a personal style is never a simple matter for a composer. When the artistic world in which he develops has extremely circumscribed expectations and seems to demand close adherence to specific conventions, it may seem almost a prohibitively difficult task, one for which nothing less than the ego of a Richard Wagner is necessary. Donizetti had no such ego, and although there are occasional references in his early correspondence to the problem of overcoming certain musical-dramatic practices that held sway over composers and public alike, we have no real reason to believe that Donizetti at this stage in his career envisioned himself as a reformer. With the advantage of considerable hindsight, Marco Bonesi, a fellow-student of Donizetti's in Bergamo, wrote many years later in his manuscript 'Note biografiche su Donizetti', dated 16 July 1861:

I participated in his first vicissitudes and at the same time shared with him my frank opinion of his music . . . Appreciating his talent more than anyone, I wanted to see him emerge with glory . . . Straightforwardly he told me that he had to cultivate the Rossinian style, according to the taste of the day. If once he made his own way a little, nothing would prevent him from developing his own style. He had many ideas how to reform the predictable situations, the sequences of introduction, cavatina, duet, trio, finale, always fashioned the same way. 'But', he added sadly, 'what can one do about the blessed theatrical conventions? Impresarios, singers, and the public as well, would hurl me into the farthest pit at least, and—*addio per sempre*.'[1]

Though this is an appealing statement, Ashbrook reminds us that Bonesi is not always a reliable witness.

In all Donizetti's published correspondence up until the period of *Anna Bolena*, there are only four letters even alluding to the problem of operatic reform:

(a) Donizetti's letter to Simone Mayr of 22 July 1822,

---

[1] Cited after Ashbrook, *Donizetti* (1982), p. 19.

concerning his *farsa*, *La lettera anonima* (Naples, Teatro del Fondo, 29 June 1822): 'I send you the newspaper article not so much to inform you of its praise as to demonstrate how much I try not to deviate from good style; and even if I lack the ability to restore music to its former glory, at least I will not have the sin of being one of its despoilers' [Z. No. 14]. This surely refers to the article of 1 July in the *Giornale del Regno delle Due Sicilie*, the key paragraph of which is quoted by Weinstock:

What pleases in the quartet is to see made fresh again the oldtime procedure of our so-called concerted pieces without those cabalettas and that symmetry of motives which obliges all the actors to repeat the same musical phrases no matter what the very different emotions agitating them may be. This is a good step toward that school of dramatic music which made the name 'Neapolitan' famous in all the theaters of Europe and to which we hope to be able to see it finally returned.[2]

Mayr would have been pleased with this judgement.

(b) Donizetti's letter to Simone Mayr of 15 June 1826, reporting his work on *Gabriella di Vergy*: 'Dearest maestro, I am setting to music for my own pleasure the *Gabriella [di Vergy]* of Carafa;[3] I know that the music [of his opera] is beautiful, but what could I do, I was seized by the desire, and now I am satisfying myself fully' [Z. No. 27]. *Gabriella di Vergy* was not written on commission; indeed, although Donizetti revised it in the hopes of a production in 1838, the work was not performed until after his death. What seems to have been important for Donizetti in this libretto were its violent situations and tragic conclusion, rarities in librettos of the period despite the example of Rossini's *Otello* (1816). No critic has demonstrated, however, that there are significant musical innovations in *Gabriella*.[4]

(c) Donizetti's letter to Mayr of 2 February 1828, describing the first-act Finale of his *opera seria*, *L'esule di Roma* (Naples, Teatro San Carlo, 1 January 1828):

The Finale of the *Proscritto* [i.e., *L'esule di Roma*], which is a Trio, is very

---

[2] Weinstock, op. cit., p. 34.

[3] The libretto of *Gabriella di Vergy* was written by Andrea Leone Tottola for the Neapolitan composer Michele Carafa. It was first performed in Naples, at the Teatro del Fondo, on 3 July 1816.

[4] See the study by Don White, 'Donizetti and the Three Gabriellas', *Opera*, xxxix (1978), 962–70. Ashbrook, *Donizetti* (1982), refers only to aspects of the libretto and distribution of voices; see pp. 39–40 and 297–8.

effective on stage. . . . Next year I will finish the first act with a Quartet and the second with a death scene written in my own fashion.[5] I want to cast aside the yoke of the finales . . . but for now I will never again close with a Trio, since everyone tells me that even should I die, return to the womb of Signora Domenica [his mother], and be born again, I would never do anything to match it . . . but I, encouraged, feel myself capable of even finer things. [Z. No. 38]

Though this is certainly an important statement, one must not overvalue it. A certain mode of constructing internal finales did indeed dominate Italian opera, but Donizetti hardly abandons it after *L'esule di Roma*. For all their effectiveness, finales in *Anna Bolena*, *Parisina*, *Gemma di Vergy*, *Lucia di Lammermoor*, and *Belisario*, to name only a few examples, adhere closely to this archetype. What is more, the Terzetto Finale of Act I of *L'esule di Roma* is an equally conventional example of trio construction, complete with a final cabaletta in which all three characters sing the entire main theme in turn (Murena, for considerations of range, in the dominant), followed by a short transition and a repetition of the theme, mercifully sung only once *a 3*. It is a good tune, to be sure, but Donizetti is scarcely avoiding 'those cabalettas and that symmetry of motives' for which the reviewer of *La lettera anonima* praised him in 1822.

(d) Donizetti's letter to his father of 4 May 1830, concerning his *azione tragico-sacra*, *Il diluvio universale* (Naples, Teatro San Carlo, 28 February 1830): 'I dare say that I worked very hard on this music, and I am pleased with it. If you expect to find cabalettas, don't even try to hear it, but if you seek to understand how I attempt to divide sacred from profane music, then endure it, hear it, and if it doesn't please you, whistle' [Z. No. 55]. This implies not a distaste for cabalettas in general, but a belief that they may be inappropriate for a sacred style. It would be fascinating to explore the extent to which Donizetti's intention is realized in the opera. Still, one need go no further than Rossini's *Mosè in Egitto* (and even more, as Ashbrook points out, its French revision, *Moïse*) to find the model for a major character whose

---

[5] *Gabriella di Vergy*, though unperformed, had concluded with a particularly grisly death scene. During the next few years, Donizetti was to live up to his promise: *Il paria* (1829) and *Imelda de' Lambertazzi* (1830) both preceded *Anna Bolena* in featuring tragic endings.

bearing is sacred and noble even though he is surrounded by typical operatic heroes and heroines.[6]

These quotations from Donizetti's correspondence show he was cognizant of having been born into an operatic world defined by the imposing presence of Rossini, but one can hardly read into them a picture of Donizetti the reformer striving to free himself from oppressive chains. It is unfortunately very difficult to evaluate these statements fully, for we cannot pretend to know well Donizetti's earlier operas. There are no complete modern editions of these works, though a few are available in reductions for piano and voice. Even in those cases, textual problems, authentic and non-authentic revisions, stand in our way. It is frustrating to analyze a piece from *L'esule di Roma* only to discover it was added for a revival of the opera more than a decade after its 1828 première. Ultimately we need to trace the development of Donizetti's approach to various musical forms and melodic styles, but for now we can begin to discover some of the aesthetic issues that concerned him. Instead of examining his words or abstractly analyzing his operas, we can catch him in the act of composition. Peppered throughout the autograph of *Anna Bolena* are numerous places in which the composer reconsiders music already written, replaces unwanted passages, and alters the structure of phrases, sections, or even entire pieces. Analysis of these passages can assist us in penetrating some important elements of Donizetti's art in *Anna Bolena* and in assessing its position among his operas.

[6] Ashbrook, *Donizetti* (1982), p. 228, makes the fascinating point that when Donizetti 'revised *Il diluvio* for its production at Genoa in 1834 he made the score sound more Italian and less French by inserting a number of cabalettas to round out arias that originally had none.'

IV. Synopsis and Overview

Anna Bolena is the second of four operas by Donizetti exploring the personalities and passions of the English Renaissance Court. The other three are dominated by the figure of Anne Boleyn's daughter, the future Queen Elizabeth I: *Il castello di Kenilworth* (1829), *Maria Stuarda* (1834), and *Roberto Devereux* (1837). In *Anna Bolena*, Romani and Donizetti brought to the stage the major protagonists in the story of Henry VIII's second marriage: Anne, her brother Lord Rochford, her early suitor Henry (here Richard) Percy, the courtier Mark Smeaton, her lady-in-waiting, future wife to the King, Jane Seymour, and of course Henry VIII himself, embodiment of the absolute monarch. Although the libretto is based on actual events, it makes no pretense to historical accuracy. Indeed, in an introductory paragraph to the libretto Romani wrote:

Henry VIII, King of England, enamoured of Anne Boleyn, repudiated Catherine of Aragon, his first wife, and married Anne; but very soon, having developed an aversion to her and becoming attracted to Jane Seymour, he sought excuses to end his second marriage. Anne was accused of having been unfaithful, and among her accomplices were cited her brother Count Rochford, the court singer Smeaton, and other courtiers. Smeaton alone confessed his guilt; and on the basis of this confession Anne was condemned to death together with all the other accused. It remains uncertain whether she really was guilty. The deceitful and cruel spirit of Henry VIII suggests rather that she was innocent. The author of this Melodrama has subscribed to that belief, more appropriate for a theatrical work; as a result, may he be pardoned if in some details he has deviated from History.

Though Romani's tone suggests he faced History unaided in writing *Anna Bolena*, we know this to be untrue. His sources, as usual drawn from contemporary theater or earlier operatic librettos, included both the drama *Enrico VIII, ossia Anna Bolena* (Turin, 1816) by Ippolito Pindemonte, itself a translation of

Marie-Joseph de Chenier's *Henri VIII* (Paris, 1791),[1] and the *Anna Bolena* of Count Alessandro Pepoli (Venice, 1788).[2]

Anna Bolena is divided into two acts. In the first, each leading character is introduced, Enrico sets his trap for Anna, and she is caught in it. The second act develops the emotions of the characters, making room for two powerful dramatic confrontations: Giovanna's admission to Anna that she is her rival, and Percy and Anna's revelation to Enrico that they were betrothed before her marriage to the King. It concludes with the death of Anna and her 'accomplices'. Within this framework, the opera is articulated as a series of discrete musical compositions, listed in Table I (see p. 22). The titles follow the precise indications or suggestions of Donizetti's autograph.

Three methods combine to help define the boundaries of each composition. First, the structure of the manuscript can be useful. New compositions always begin with new gatherings, the standard unit in the autograph of *Anna Bolena* being a gathering of two bifolios. Frequently Donizetti numbers the gatherings within a composition. He writes '1' in the upper-left-hand corner of the first folio of the first gathering, and then proceeds progressively throughout the composition. Only in the case of the Introduzione (No. 1) does his enumeration of gatherings contradict the indications of Table I, for here he began numbering anew in the middle of the composition. Surely this is a consequence of the extensive compositional changes that preceded the definitive form of the Introduzione, changes that will be discussed in Chapter XII. Second, Donizetti frequently supplies titles for individual pieces, either writing them at the start of a composition or using phrases such as 'Segue Aria Percy' or 'Dopo il Terzetto'. Though these indications can occasionally be inconsistent, they are normally an invaluable guide to the way Donizetti conceived the structure of his opera. Finally, most of

---

[1] See Franca Cella, 'Indagini sulle fonti francesi dei libretti di Gaetano Donizetti', *Contributi dell'Istituto di Filologia Moderna*, iv (1966), 343–584, in particular pp. 440–7.

[2] Pepoli was the grand-uncle of Count Carlo Pepoli, author of the libretto for Bellini's *I Puritani*. An eccentric nobleman, he constructed a private theater in Venice, the Teatro di S. Vitale, where his own plays and operas could be performed, operated his own printing shop 'dalla Tipografia Pepoliana' to publish his texts, and frequently acted and sang in these productions.

Table I. The formal numbers in *Anna Bolena*

Piece	Key
Sinfonia	D major
Act 1	
1. Introduzione	E♭ major
2. Recitativo Dopo l'Introduzione e Duetto [Giovanna, Enrico]	D major
3. Recitativo e Cavatina Percy	B♭ minor; E♭ major**
4. [Recitativo] Dopo la Cavatina di Percy Quintetto	C major
5. Scena e Finale 1°	
(a) Scena [e Cavatina] Smeton	A♭ major
(b) Duetto finale [Anna, Percy]*	B♭ major
(c) [Finale 1°]	D major
Act II	
6. Introduzione	G major; to A minor
7. [Recitativo] Dopo l'Introduzione Duetto [Anna, Giovanna]	C major
8. Coro, [Scena] e Terzetto [Anna, Percy, Enrico]	
(a) Coro	E♭ major
(b) Scena e Terzetto	C major
9. [Recitativo] Dopo il Terzetto Aria [Giovanna]	E major
10. [Recitativo] Aria Percy	A major**
11. Ultima Scena [Anna]	
(a) [Coro]	F minor
(b) [Scena ed Aria Finale Anna]	F major; G major; A♭ major; E♭ major

* 5(b)′ Duetto nel Finale dell'Atto 1ᵐᵒ rifatto   B♭ major
** These numbers are transposed in the piano-vocal scores.

the compositions in *Anna Bolena* are tonally closed, that is, their first and last lyrical sections are in the same key. There is some evidence that one of the pieces contradicting this procedure, the Cavatina of Percy (No. 3), was originally planned to be tonally closed (see Chapter VIII).

Only in the cases of Finale I (No. 5) and the Introduzione of Act II (No. 6) are firm boundaries between numbers attenuated. In the Finale, a Cavatina for Smeton and a Duetto for Anna and Percy precede the four-part Finale 'proper', whose second and fourth parts are the expected Largo and concluding *stretta*. Although the Cavatina, Duetto, and Finale 'proper' each has its own tonal center (A♭ major, B♭ major, and D major, respectively), Donizetti provided transitional measures between them to ensure continuity. The Introduzione of Act II (No. 6), on the other hand, is essentially a woman's chorus in G major, followed immediately by recitative with arioso periods for Anna, Hervey (an official of the court), and the chorus. This section serves as a transition to the following scene between Anna and Giovanna.

Perhaps because formal structures in Italian opera of the first half of the nineteenth century appear on the surface to be so regular, no adequate treatment of them exists in the secondary literature.[3] This is not an appropriate place to present a general *Formenlehre* for the period; rather, individual models will be developed within this study as they prove necessary. It is important to state explicitly, however, that the nature of the evidence found within the autograph of *Anna Bolena* leads us inevitably to concentrate our attention on the major formal numbers, where traces of Donizetti's compositional process are most apparent. Though we will consequently seem to be slighting scenes that take place in recitative, it must be understood that one of Donizetti's principal achievements in *Anna Bolena* lies precisely in his handling of these recitatives, with their frequent passages in arioso, their rich orchestral coloring, and their intensely dramatic confrontations. Yet these are among the cleanest pages in the autograph, demonstrating an extraordinary confidence of approach, an ability to set dramatic dialogue that helps account for Donizetti's success as a composer of 'Romantic' operatic subjects. Analysis of his technique and comparison with that adopted in earlier *opera seria* by Rossini or contemporary works by Bellini would make a fascinating study in its own right.

As examples will be drawn freely from diverse numbers of the opera, we offer the following brief description and synopsis of *Anna Bolena* to help orientate the reader.

---

[3] The most important general studies in this field are Friedrich Lippmann, *Vincenzo Bellini und die italienische Opera Seria seiner Zeit*, published as *Analecta Musicologica*, vi (1969), and Julian Budden, *The Operas of Verdi*, particularly vols. i (Oxford, 1973) and ii (Oxford, 1979).

Sinfonia. The overture is a complete orchestral Allegro in a modified sonata form, with an introductory Allegro and Larghetto. It has unmistakable roots in the archetypical Rossini overture, and is very different from the 'pastiche' overtures Bellini was writing for operas such as *I Capuleti e i Montecchi* (1830) and *Norma* (1831). Noteworthy is the absence of a reprise of the first theme in the tonic; it is replaced by a short modulating, developmental passage based on that theme. Though the overture has no overt dramatic links to the opera, two of its themes do recur there: its crescendo theme opens the *tempo di mezzo* in Giovanna's Aria (No. 9);[4] the first measures of its second theme recur as the melody Anna 'hears' when, early in the opening recitative of her Ultima Scena (No. 11(b)), she thinks of her lost happiness with Percy.

1. *Introduzione.* The Introduzione is an extended opening number. Traditional examples of the genre, in which it was rare for a prima donna to participate, had featured an opening chorus, a *cantabile* movement for one or more characters, and a concluding cabaletta. Donizetti's Introduzione to *Anna Bolena* is a much more extended composition, drawing on such Rossinian models as *La Cenerentola* or *Semiramide*. Yet, while establishing at once the emotional world of the opera and introducing us to several of the central characters, it still preserves the three traditional main parts: an opening chorus, *cantabile* movement, and cabaletta, the last two both sung by Anna.

The opening scene takes place in a room of Windsor Castle, near the apartments of the Queen. The chorus laments that the King has not yet come to see his lady. They fear Anna's star has faded and that Enrico loves another ('Né venne il Re?'; E♭ major). Giovanna Seymour enters. In a short lyrical aside, she reveals herself to be Enrico's new love ('Ella di me, sollecita'; A♭ major). Barely able to face her mistress, she prays either to be freed from remorse or to love Enrico no more. Anna herself appears, obviously disturbed. At her request, the page Smeton,

---

[4] The term *tempo di mezzo*, employed frequently by 19th-century writers such as Abramo Basevi, in his *Studio sulle opere di Giuseppe Verdi* (Florence, 1859), refers to the section of music, usually in dramatic dialogue, that leads from the opening *cantabile* of an aria to its concluding cabaletta. In ensembles, the *tempo di mezzo* is normally the third section, between a *cantabile* second movement and the concluding cabaletta.

accompanied by the harp, offers two strophes of a formal song ('Deh! non voler costringere'; E♭ major).[5] The melancholy state of the Queen, he says, is in itself beautiful. One might think she were sighing for her first love. As he begins a third strophe with the same idea, 'Quel primo amor', Anna breaks in. These words have struck her heart (*cantabile*, 'Come, innocente giovane'; G major). With all her vain splendor, she muses, she would have been happier had she not abandoned her first love. But the night advances, dawn is near, and the King is clearly not coming (*tempo di mezzo*). When Giovanna asks Anna to explain what troubles her, Anna replies that she must keep hidden the reason for her sadness (cabaletta, 'Non v'ha sguardo'; E♭ major).

2. *Recitativo; Duetto (Giovanna and Enrico)*. Giovanna is left alone, torn between her love of the King and remorse towards Anna. Through a secret door Enrico enters. She tells him they must no longer see one another, but he refuses to accept this. For Giovanna, though, his love is insufficient; she must also preserve her honor. Their Duetto, which follows immediately, is in the standard four parts: an opening section with parallel formal statements for each character; a *cantabile* movement; a *tempo di mezzo*, in which the action is advanced; and a final cabaletta.[6] In the first section, Enrico assures her she will obtain everything she desires ('Fama! sì: l'avrete'; D major). Yet they cannot marry, Giovanna replies, for the altar is banned to her. Enrico accuses her of loving only the throne. As Giovanna protests, he tells her that Anna too offered him love but really sought only the throne (*cantabile*, 'Anna pure amor m'offria'; C major). Offended, Giovanna reminds him that she did not freely offer him her love; the King stole it from her. As she begins to depart, Enrico stops her (*tempo di mezzo*). He will marry her, while Anna will be punished for the worst of sins: giving him a heart that was not hers to give. When Giovanna asks how he means to break their ties, Enrico refuses to say. Do not use cruel means, she begs him (cabaletta, 'Ah! qual sia cercar non oso'; D major); do not fill me

---

[5] It is hard not to draw the parallel with Cherubino and the Countess in Act II of Mozart's *Le nozze di Figaro*.

[6] For an extended discussion of the Duetto form in Italian opera of the 19th century, see my article, 'Verdi, Ghislanzoni, and *Aida*: The Uses of Convention', *Critical Inquiry*, i (1974), 291–334.

with more remorse than I feel already. Enrico, not responding to this plea, simply assures her again of his love.

3. *Recitativo; Cavatina Percy*. Lord Rochefort, brother to the Queen, is standing in a park outside Windsor Castle. The exiled Riccardo Percy enters, and they embrace. Percy, who has been called back by the King, intends to present himself to his sovereign as the latter departs for the hunt. Yet his return is bitter, for he still loves Anna. Rochefort reports that Anna's sole joy is being Queen. When Percy refers to his continued love for her, Rochefort urges him to be still, but Percy cannot control his emotions. He expresses them in his Cavatina, which in this period normally means 'entrance aria'. It is constructed in the standard two sections, a *cantabile* and a cabaletta, joined by a *tempo di mezzo*. He recalls his exile (*cantabile*, 'Da quel dì che, lei perduta'; B♭ minor). Though all life seemed a tomb, he was sustained by thoughts of Anna. Sounds of preparation for the hunt are heard, as pages, hunters, and servants enter (*tempo di mezzo*). Percy asks Rochefort whether Anna will appear, but Rochefort, fearful for both his friend and his sister, does not answer. Percy is overwhelmed with rapture at the thought that he will see her once again (cabaletta, 'Ah! così nei dì ridenti'; E♭ major).

4. *Recitativo; Quintetto*. The King and his courtiers appear, as do Anna and her ladies. When Enrico questions her presence, she explains that she has come because he has been absent for so many days. With an undercurrent of threat, he responds that she has been constantly in his thoughts. Seeing Percy, Enrico calls him forth. As Percy bows to kiss the King's hand, Enrico points to Anna: it was she who convinced him of Percy's innocence. This is too much for Percy, who turns with passion to her. The Quintetto follows immediately, a four-part ensemble with two main lyrical poles (a *cantabile* section and a final cabaletta or *stretta*), each preceded by a section in which action occurs (a *tempo d'attacco*[7] at the outset and a *tempo di mezzo* before the cabaletta).

---

[7] *Tempo d'attacco* is a somewhat less common term in the period: Basevi does employ it frequently. It is often more useful for descriptions of Verdi's music than for earlier composers, since the latter usually provide quite extensive musical and dramatic tableaux at the beginning of an ensemble. If we understand a *tempo d'attacco* to refer to a short introductory section of an ensemble, one that does little more than prepare the second, *cantabile* section, then this first section of the Quintetto is a good example of the genre, closer to many Verdian compositions than to those of Rossini.

In the *tempo d'attacco*, Percy expresses his gratitude to Anna ('Voi, Regina!'; C major). As Enrico watches, Percy falls to his knees and kisses Anna's hand. The Queen is also deeply moved (*cantabile*, 'Io sentii sulla mia mano'; A♭ major). Rochefort tries to calm Percy, but the latter is obsessed by the thought that Anna continued to be concerned about him during his exile. Enrico instructs Hervey to follow their every thought and motion. Full of apparent goodwill, Enrico tells Percy he must remain at court (*tempo di mezzo*), then bids farewell to Anna and prepares his departure for the hunt. Each character comments on this auspicious beginning of the day, yet expresses fears as to what may yet occur (*stretta*, 'Questo dì per voi spuntato'; C major).

5. *Scena; Finale I°*. The standard Finale design in this period also involved a four-part ensemble, similar to the Quintetto just described. Usually the Finale was further extended, however, either by adding other internal sections or by expanding it at the start, the technique Donizetti employed in *Anna Bolena*. He prefaced the four parts of the Finale 'proper' with a Scena e Cavatina for Smeton (with an abbreviated *cantabile* and full cabaletta) and a Scena e Duetto for Anna and Percy (a two-part Duetto, consisting only of the opening section, parallel strophes for the two characters and a short dialogue, followed by a full cabaletta). Recitative with extensive arioso passages leads directly from one element of this design to the next.

Smeton has penetrated secretly to a room near Anna's apartment. One day, called there to give a private concert, he had taken a small portrait of his mistress. Ever since he has worn it at his breast. Fearing that his passion may be discovered, he now seeks to return the portrait. First he must kiss it again (*cantabile*, 'Un bacio, un bacio ancora'; A♭ major). To the portrait he reveals the passion he must hide from his mistress (cabaletta, 'Ah! parea che per incanto'; A♭ major). Suddenly there is a noise, and Smeton hides behind a curtain. Anna and Rochefort enter. He begs her to receive Percy, insisting that the risk she faces is graver if she refuses. Her resistance crumbles, but she warns Rochefort to stand guard. When Percy appears, she asks whether he wishes to upbraid her for betraying their oaths. Yet, she has made amends: her crown is of thorns. Seeing her unhappiness, Percy responds, his anger dissolves. For him she remains Anna, and even if the King despises her, Riccardo continues to love her (opening section of the Duetto, 'S'ei t'abborre, io t'amo ancora'; B♭ major). Anna insists that her marriage bonds are sacred. Percy must never again speak of love. She urges him to have pity on her and to leave England, but Percy answers that he must remain near her or die (cabaletta, 'Per pietà del mio spavento'; B♭ major). When Anna announces they must never meet again, Percy draws his sword, as if to kill himself. The Queen screams, and the terrified Smeton emerges from his hiding place. At this, Anna swoons. Rochefort enters to reveal, too late, that the King has returned. Enrico, surveying the situation, says that in Anna's face he can read her infidelity (opening section of the Finale 'proper', 'Tace ognuno'; D major). Smeton defends his mistress, when suddenly her portrait falls from his breast. To Enrico this is further proof of her guilt, and even Percy believes Smeton is his rival. Anna begs the King for a moment to recover her senses (*cantabile*, 'In quegli sguardi impresso'; D major). Enrico tells her it would be best if she died at once and orders them to separate cells (*tempo di mezzo*). Though Anna tries to speak, he refuses to listen: explain yourself, he fulminates, to your judges. 'Judges', repeats Anna, and in despair envisions her certain death (*stretta*, 'Ah! segnata è la mia sorte'; D major). She tries to catch Enrico's attention, but he strides off stage, leaving Anna, Smeton, Percy, and Rochefort to the guards. Giovanna Seymour looks on horrified at the turn of events.

6. *Introduzione*. The second act opens outside the rooms where Anna is being kept prisoner. In a short chorus, her ladies-in-waiting lament that Anna's friends have now abandoned her ('Oh! dove mai ne andarono'; G major). As Anna appears, they try to comfort her. Hervey announces that Anna's handmaidens must testify before the Council. She bids them go, entreating them to bear witness to her innocence.

7. *Recitativo; Duetto* (*Anna and Giovanna*). Anna falls on her knees in prayer, as Giovanna enters with a proposal from Enrico. Though the King insists their marriage be annulled, he will spare her life if Anna admits her guilt. When the Queen refuses thus to stain her honor, Giovanna presses her: not only the King begs her to accept, but also the unfortunate woman Enrico has destined for the throne. 'Who is she?' asks Anna, pronouncing a terrible curse on her ('Sul suo capo aggravi un Dio'; C major). This is the first section of a Duetto that in many respects is structurally unusual.

Rather than featuring parallel strophes for the two characters, its first section centers largely on Anna's invective. Her curses grow more violent, until Giovanna, who cowers before her, throws herself at the Queen's feet. As Anna realizes that Giovanna herself is the hated rival, Giovanna begs for pardon (*cantabile*, 'Dal mio cor punita io sono'; G major), while Anna is reduced to isolated cries of 'va!' or 'fugge!' Enrico seduced her, Giovanna explains, and although she does love him, that love is her shame. Moved by Giovanna's sorrow, Anna raises her up and grants her forgiveness (cabaletta, 'Va, infelice, e teco reca'; A minor, C major). Anna's pardon, says Giovanna in a period entirely different from the Queen's, is worse than the wrath she deserves.

8. *Coro, Scena e Terzetto.* In a vestibule outside the Council chamber the courtiers await news ('Ebben? dinanzi ai giudici'; E♭ major). They fear that Smeton, his inexpert heart seduced by false hopes, may have confessed some misdeed. Hervey appears, calling for Anna and Percy to come before the judges; Smeton has indeed confessed and Anna is lost. The courtiers disappear as Enrico emerges, ordering Smeton back to his cell so that the Page will think he has saved his mistress until it is too late. When Anna and Percy enter, Enrico tries to depart, but the Queen stops him, begging for death rather than dishonor. He scorns her plea, taunting her for having descended to the level of a Percy. Yet, says the latter, you did not disdain to make me your rival and steal her from me. He swears the Queen is innocent. If she resisted you, the King replies, she did not resist her Page, who has confessed. At that slander Anna regains her stature, accusing Enrico of having tricked Smeton. Her only crime was preferring the crown to the heart of Percy. The King cuts the rapturous Percy short: you will both die, he says, beginning the first section of this standard four-part ensemble ('Ambo morrete, o perfidi'; C major). Justice, he continues, died in this court when a Queen (Catherine of Aragon) was compelled to descend from her rightful throne for Anna, but it will speak again. If anyone has a right to revenge, Percy interjects, it is he, for the King is the one who has profaned a marriage vow: Percy and Anna were once betrothed. From their earliest years, they were meant for one another (*cantabile*, 'Fin nell'età più tenera'; G minor). Percy will restore her life and honor, the life and honor Enrico stole away.

Anna reflects that the throne has brought her only sorrow, while Enrico threatens revenge. He insists they be taken before the Council (*tempo di mezzo*), where they can confess their betrothal. A worse fury still will descend on their heads. Another woman will ascend the English throne, one more worthy of it, and Anna's name will be covered with infamy (cabaletta, 'Salirà d'Inghilterra sul trono'; C major). Percy and Anna pray that England may never know the horrid sufferings of Anna. They are led off by the guards.

9. *Recitativo; Aria Giovanna.* Enrico, left alone, ruminates over the news that Anna and Percy were formerly betrothed. Perhaps it is merely a scheme to allow her to escape punishment as his guilty wife. Even so, the law will descend on her, also dragging her daughter's reputation into ruin. A repentant Giovanna enters. She wants to leave the court forever, hoping thereby to save Anna's life. The King assures her he will hate Anna even more if Giovanna's love is extinguished. Giovanna, to her shame, replies that her love remains strong (*cantabile*, 'Per questa fiamma indomita'; E major). Hervey returns (*tempo di mezzo*), announcing that the Council has dissolved the royal marriage and condemned Anna and her accomplices to death. The chorus, restating the music of No. 6 (Introduzione, Act II), declares that their only hope lies in royal clemency. Giovanna adds her tears, but Enrico will not listen. Justice, he affirms, is the primary virtue of a King, and he signs the death-warrant. Giovanna reminds him that heaven and earth are watching (cabaletta, 'Ah! Pensate che rivolti'; E major). As the chorus joins her plea, Enrico summons the Court again. Giovanna's Aria, with its two standard lyrical sections, is distinguished by a relatively extensive *tempo di mezzo*.

10. *Recitativo; Aria Percy.* In the prisons of the Tower of London, Percy and Rochefort learn that both have been condemned to death. Hervey enters to announce that the King has pardoned them. When Percy is told that Anna must still die, however, he refuses Enrico's pardon. Rochefort does the same, despite Percy's impassioned plea that he should live (*cantabile*, 'Vivi tu, te ne scongiuro'; A major). Hervey orders them separated (*tempo di mezzo*), and they take their last farewells. They will meet again in death, says Percy (cabaletta, 'Nel veder la tua costanza'; A major). In this, his last hour, he leaves nothing behind and has no regrets. Percy's Aria is cast in the same mold as

the earlier ones in the opera. The two friends are escorted out by soldiers.

11. *Ultima Scena Anna.* The woman's chorus describes Anna's sorrowful state ('Chi può vederla'; F minor). The Queen enters, delirious. Why are they crying on the day of her wedding? The King awaits. Don't let Percy know, for she must hide from his glances. He accuses her. She begs forgiveness. But will he forgive her? This section of dramatic recitation is dominated by constantly shifting lyrical periods and orchestral interjections. Guide me to my native castle, she muses, to recapture at least a single day of our early years of love (*cantabile*, 'Al dolce guidami'; F major). To the accompaniment of a solemn march, Percy, Rochefort, and Smeton are brought forth. Anna, in her delirium, asks the crestfallen Smeton to play a sad song for her, one expressing her broken heart. She herself sings the song, Donizetti's arrangement of 'Home, sweet home' ('Cielo, a' miei lunghi spasimi'; G major). As festive sounds are heard, Anna regains her senses. The people are acclaiming their new Queen, she realizes, and Anna must now be sacrificed. With terrible dignity she refuses to curse Enrico and Giovanna, preferring to descend to her tomb with words of pardon and pity on her lips (cabaletta, 'Coppia iniqua'; E♭ major). As she swoons, Smeton, Percy, and Rochefort are escorted to their death, and the curtain falls.

This Ultima Scena is the most impressive example in the opera of Donizetti's treatment of inherited forms. None the less, it is essentially a 'Gran Scena' of the type found in many of Rossini's operas, involving several full cantabile sections, linked by choral interventions and recitative, and concluding with a complete cabaletta.[8] It surpasses earlier examples by the quality of the dramatic recitation, the extraordinary way Anna wavers between sanity and madness, and the sense of dramatic and musical unity the entire scene conveys. Donizetti would look back to it when he

[8] Both Elisabetta's 'Fellon, la pena avrai' and Armida's 'Se al mio crudel tormento' are large-scale concluding arias with three separate lyrical sections; Ermione's 'Gran Scena' includes at least four closed sections, beginning with 'Dì, che vedesti piangere'. These are from Rossini's *Elisabetta, Regina d'Inghilterra* (1815), *Armida* (1817), and *Ermione* (1819), respectively. And of course it is not only to Rossini we must look for such pieces but, perhaps even more important, to Donizetti's teacher, Mayr. The 'Scena dei Solitari' in *Ginevra di Scozia* for Ariodante is a well-known example. Donizetti may make the genre his own here, but he did not invent it.

came to write *Lucia di Lammermoor* in 1835, and Bellini would remember it too for *I Puritani*. Imogene's mad scene in *Il pirata*, however, written by Bellini three years earlier, could not have been far from Donizetti's mind. Certainly Anna's 'Coppia iniqua' and Imogene's 'Oh! sole ti vela' are cut from the same cloth:

Ex. 2. A comparison of themes from Bellini's *Il pirata* and Donizetti's *Anna Bolena*

There is no trick in the arsenal of a prima donna that Donizetti did not employ in this finale, yet each emerges from the dramatic situation: intense declamation for her recitative; sustained lyrical singing as she remembers the happy past; tender arioso phrases as she asks her page to play a melody 'like the sigh of a dying heart'; the simple, folk-like prayer based on 'Home, sweet home' (it is difficult for Anglo-Saxon audiences to ignore the reference, though despite it the naïve emotion of the piece is touching); and finally the violent cabaletta, with its melodic line soaring and swooping from one register to another. Her words tell us that she will not cry for revenge in her hour of death; her music belies them.

As in the case of the dramatic recitative of *Anna Bolena*, the Ultima Scena is remarkably free of compositional alterations in Donizetti's autograph. Hence, it will not figure prominently in this study. Let me reiterate that the following pages make no pretence to being a complete analysis of *Anna Bolena*: they represent an attempt to learn as much as possible about the work from a study of Donizetti's compositional process.

As a drama, *Anna Bolena* is very well constructed: the various elements of the plot are clearly developed and the characters are coherent personalities. The numbers all stem from precise archetypes, well established in the tradition. Yet as we penetrate further into the work, we become ever more aware of the ways in which Donizetti sought to impose on that tradition his own musical personality. Much of Donizetti's greatness as a composer, and not only in *Anna Bolena*, lies precisely in the extent to which he was successful.

V. Melodic Detail

ON the most intimate level, we can watch Donizetti refine his melodic ideas. It is characteristic of the composer's mature style that instead of simply repeating melodic phrases and cadences, he explicitly varies them. In the music of Rossini melodic variations of this kind were less frequently notated, although singers were expected to provide them in addition to their normal practice of inserting cadenzas at appropriate moments and ornamenting the repeat of a cabaletta theme.[1] Many passages in *Anna Bolena*, however, were first conceived either without melodic variants or with less extensive ones. The simplest type of example is a short, essentially repeated, cadential phrase, such as the one given in Example 3 from the end of Anna's solo period in the first section of her Duetto with Giovanna (No. 7):

Ex. 3. An altered cadence for Anna in her Duetto with Giovanna (No. 7)

---

[1] On many occasions Rossini himself prepared sets of variants for singers. A superb set for Ninetta's Cavatina, 'Di piacer mi balza il cor' in *La gazza ladra* is printed as Appendix IC in the critical edition of that opera, ed. Alberto Zedda and published by the Fondazione Rossini (Pesaro, 1979), p. 1080.

The two-measure cadence is repeated with no change at all in the orchestral accompaniment (indeed, Donizetti does not bother to write the parts out a second time), but even in the original layer there is a change in the melody at the fourth measure, one that increases the momentum of the vocal line. Donizetti's subsequent alteration in the first measure of the repeat intensifies further this momentum, giving Anna's curse greater dramatic power.

A similar situation occurs in a two-measure repeated phrase in the last measures of the Introduzione (No. 1). Again a small change in the second measure of the original layer in Example 4(a) is followed by a massive revision of the first measure:

Ex. 4. Cadential revisions in Anna's cabaletta in the Introduzione (No. 1)

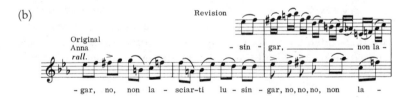

* Because of the poor declamation in the next measure, Donizetti altered the text underlay of these measures in the original version.

The extent to which Donizetti's revisions can alter the whole character of a cadential phrase is apparent in the set of cadences concluding Anna's cabaletta theme in this same Introduzione (Example 4(b)). The original layer provided already some variation in the two-measure repeated cadence, finding in the accented lower auxiliary on the second beat in the first measure a figure that could be repeated several times. Yet nothing in that original layer prepares us for the cascade of similar figures that Donizetti employs in his revision. Where the original is rhythmically flat, depending on a lengthy series of eighth notes unbroken until the penultimate measure of the example, the revision stresses rhythmic variety. After the pairs of sixteenth notes, Donizetti retreats to quarter notes, declaimed on a single pitch, in order to prepare the octave leap, the descending chromatic scale, the arpeggio upwards, the accented high C, and the final arpeggio down to the tonic that transform these final measures of the cadence into a prima donna's delight.

Every revision does not necessarily aim for greater virtuosity. In the cabaletta theme of Anna's Ultima Scena (No. 11), the original cadences are both longer and more elaborate than the revised ones, specifically inviting the singer to provide a cadenza by means of the fermata (see Example 5).

The final five measures are a varied expansion of the previous two-measure cadence. Given the dramatic situation and the vehemence of the entire cabaletta theme, Donizetti's decision to avoid any weakening of the rhythmic intensity here was unquestionably correct, although his fourfold repetition of the same rhythmic figure was not a particularly inventive musical solution. It is striking that the revision extends the range of the melodic line once again to the high C. In fact there are very few high Cs notated in the part of Anna, and all but one are the result

Ex. 5. A cadential revision in Anna's cabaletta in the Ultima Scena
(No. 11)

of Donizetti's revisions. This suggests strongly that at first he may
have thought Pasta was uncomfortable with this pitch; perhaps,
indeed, some of these revisions were carried out in collaboration
with the singer.

Cadential changes do not only affect Anna's part, however, nor
are they necessarily so elaborate. In a Duetto, a twofold cadence
frequently features one singer the first time and the other at the
repetition. In Example 6, the final cadences of the *cantabile* section
in the Duetto for Giovanna and Enrico (No. 2), Enrico's phrase
originally began with an arpeggiated melodic figure parallel to
that of Giovanna's three bars later:

Ex. 6. A cadential revision in the *cantabile* of the Duetto (No. 2)

Donizetti's revision was minor, changing only details of the figuration, not its essential thrust, yet the series of lower auxiliary notes for Enrico is sufficient to give a slightly different musical tint to his character.

In notating variations within cadential figures, Donizetti sometimes offers little more than what an intelligent singer might have been expected to provide. The situation is quite different for alterations he makes to melodic lines. Here fundamental aspects of his musical style are involved, and in many of the changes we can observe his ear at work, testing musical patterns, searching for that succession of pitches that appeals most to his personal taste. The cabaletta theme of the Duetto Finale for Anna and Percy (No. 5(b)) originally began as in Example 7:

Ex. 7. The cabaletta theme of the Duetto (No. 5(b))

The composer wrote out the entire cabaletta with this theme, for it recurs in the original form both at Percy's statement of the theme and at the reprise, where it is sung by Anna with an accompanying line added for Percy. There are many ways in which the revision is a distinct improvement. It handles the upper register better, providing the E♭ as a point of continuity leading to the F. (The theme eventually ascends at its climax to the G, from where it returns at the cadence to the tonic.) It helps to ease the parallelism in the second and fourth measures of the theme, avoiding the simple repetition of G–F at the conclusion of the two phrases. It also presents a far more effective declamation of the word 'spavento' (fear). Indeed, Donizetti made his melodic revision before orchestrating the piece.[2] The harmony he ultimately uses to underline 'spavento' on the fourth beat of the first complete measure in Example 7 (a C diminished triad over a B♭ pedal) would not have been possible in the original version, where a simpler E♭ major triad in second inversion was probably intended. Though the melodic revision is a small one, it affects considerably our perception of this cabaletta.

Between the theme of a melodic period and its cadences, composers of Italian opera in the nineteenth century frequently employed passages that sat over a pedal or moved regularly back and forth from dominant to tonic harmonies. One would almost be tempted to borrow a term from the world of commercial aviation and call these passages 'holding patterns'. Though the measures may appear to function more as 'filler' than as significant elements of the musical discourse, finding the right details for each specific situation was by no means mechanical. In

---

[2] It is clear that Donizetti often worked directly on the orchestral manuscript while composing, filling in vocal lines and bass, with some significant other voices, saving orchestration for a later stage. His letter to Mayr (Z. No. 31) of 30 September 1826, concerning *Olivo e Pasquale*, affirms this: 'Here I am working, and I am already almost through half of the second act, though without instrumentation; for now I leave it in that form for whatever alterations might be necessary, or rather that I may be called upon to make. For the moment the basic work is on paper.' From an examination of different colors and shades of ink, one can usually differentiate the various stages of composition. Rossini worked in much the same fashion, as I have sought to demonstrate in my article, 'Gioachino Rossini and the Conventions of Composition', *Acta Musicologica*, xlii (1970), 48–58.

the cabaletta theme from Smeton's Cavatina in the Finale 1° (No. 5(a)), the 'holding pattern' is a four-measure phrase, repeated, over an E♭ pedal, resolving to the tonic in the bass only at the final measure:

Ex. 8. The cabaletta of Smeton's Cavatina (No. 5(a))

*At later appearances this word is changed to "a lei".

The original version, though not particularly distinguished, is perfectly acceptable in itself. In the context of the preceding thematic phrases, however, the reason for Donizetti's dissatisfaction becomes immediately apparent. Rhythmic regularity is basic to this theme, which is constructed in two parts, a four-measure phrase cadencing in the tonic and its expanded repetition modulating to the dominant. To turn directly from this theme to the original continuation provides very little contrast: the dotted rhythms of the upbeats are identical; the repeated dotted note pairs (F–E♭) in the first and fifth measures recall clearly the similar figures in the second phrase of the theme. And the sequential nature of the first two measures of this phrase reproduces the sequential process that underlies the preceding theme. Donizetti's revision deals with all of these problems, providing in particular many elements of rhythmic novelty. The upbeat is transformed into a group of four sixteenth notes, contrasting well with the same figuration on the downbeat in the next measure. The syncopation on the D♮ is particularly appealing after a theme where accents fall with almost depressing regularity on the downbeat of each measure. In this way, Donizetti is able to sustain the melodic and rhythmic interest of the entire period.

Probably the most fascinating examples of changes in melodic detail, however, are those that transform a significant theme. There are three examples in *Anna Bolena* that are worthy of our attention. Within the Sinfonia, Donizetti had particular difficulty defining the ultimate shape of the second theme. Alterations are apparent both in its original presentation in the dominant and its reprise in the tonic. A number of different layers seem to be present, and it is difficult to reconstruct the sequence of events with any precision. At least one of these layers, however, visible in both the dominant and tonic areas, assigned the opening of the theme to a solo clarinet and divided its first eight measures into two parallel phrases of two measures each, followed by a four-measure cadence ($2\widetilde{+}2+4$). Example 9 presents the entire theme at its first appearance in the dominant:

Ex. 9. Revisions in the second theme of the Sinfonia

etc.

In the original version it is precisely the initial musical idea that returns after four measures of contrasting material to prepare the cadence of the period. Thus the period not only begins with a more symmetrical melody but also depends upon symmetrical repetition in its overall construction. Donizetti's revision eliminates all these explicit parallelisms, transforming the first four measures into a continuous musical idea, one that has only generic links to the concluding phrase of the period. The links are obviously strong enough, however, to give the entire period melodic coherence without insisting on melodic identity.

Two striking examples from Anna's part show the extent to which detailed changes of this kind can transform the quality of a melody. Her solo initiating the *stretta* of the Finale 1° (No. 5(c)) was originally conceived like this:

Ex. 10. Anna's phrase initiating the *stretta* of Finale 1° (No. 5(c))

Here mm. 9–12 were identical to mm. 1–4. The altered version, worked out before Donizetti orchestrated the passage, preserves the essential downward motion of mm. 1–4 but moves first from the opening pitch, F♯, up to G, so that the descent, when it does come, is more precipitous and hence more emphatic. A similar example occurs in Anna's *cantabile*, 'Come, innocente giovane', within the Introduzione (No. 1). In the original version, found in Example 11, the opening of the melody is more repetitive,

Ex. 11. Anna's *cantabile* in the Introduzione (No. 1)

though even here Donizetti introduces variation (turning the arpeggiated eighth notes of m. 2 into the scalar sixteenths of m. 6). By canceling the identity of mm. 1 and 5 and the near-identity of mm. 4 and 8, and by eliminating the melodic sequence of m. 3, Donizetti gives the melody the vigor the original version lacked, while expanding its range downwards to highlight Anna's lower register.

Even later in his life, Donizetti frequently notated at first simple phrases that gain their individuality only through the process of revision. To take but a single example, the orchestral theme accompanying Don Pasquale's impatient pacing at the beginning of *Don Pasquale* (1842/3) was originally conceived as follows:

Ex. 12. Opening orchestral theme in the Introduzione of *Don Pasquale*

In the autograph, also located in the Ricordi archives, Donizetti alters the strict repetitions of the first two phrases to give the melody its most characteristic elements: the subtle chromatic inflections and the sudden upward leaps, particularly effective in a tune which, apart from octave displacements, is essentially linear.

Melodic variation that disguises symmetric phrases in a period, as in the last few examples, is rare in the music of Rossini. When altering the repetition of a phrase, Rossini generally rewrites its conclusion, where the harmony may veer off the tonic to the third, fifth, or sixth degrees. In our examples, Donizetti begins with a phrase originally conceived as symmetrically as Rossinian phrases tend to be, only then to embed this symmetry into a continuous melodic flow. From a more conventional framework, he develops a personal melodic statement.

Melody is the sustaining force in much nineteenth-century Italian opera. At its shrine worship opera lovers of every description from simple dilettantes to *cognoscenti*. Donizetti's notated melodic style differs significantly from Rossini's in cadences, holding patterns, and themes. Part of these differences were the result of Donizetti making explicit certain variation techniques normally left to the singer's art by Rossini. In revising his melodic ideas to fix a specific interpretation of his music, Donizetti reduced his dependence on the whims of the interpreters and allowed his own melodic impulses to develop more freely. There is still room for improvised ornamentation in *Anna Bolena*, to be sure, but much less than in a typical Rossini *opera seria*. Our examples demonstrate the extent to which this development, significant not only for Donizetti but also for the future course of Italian opera, was won largely through the conscious elaboration of more straightforward musical ideas.

VI. The Search for Continuity

FOR certain sectors of the public, the operas of Donizetti and his contemporaries exist to show off the qualities of their favorite singers. The largely sectional structure of individual numbers becomes an invitation to stop performances time and again by roaring approval for particularly impressive vocal feats. Attempting to eliminate this practice altogether would be foolish, even quixotic, for there are unquestionably moments within the operas, usually after entire numbers, where applause is not out of place. Yet Donizetti was keenly aware of the need to keep the drama and music moving forward, to avoid the excessive use of full breaks within a musical number. That he sought this continuity, while preserving the sectional structure inherited from Rossini, can be seen from the extent to which he first conceived periods with fixed and definite boundaries, provided by orchestral introductions, cadences, etc., only to alter them and overlap sections in the autograph. Examples abound, of which those present in two numbers, the Recitativo and Duetto for Anna and Giovanna (No. 7) and the Coro, Scena, and Terzetto (No. 8), may suffice.

Let us begin with a few simple yet significant cases in the Coro, Scena, and Terzetto. At the very instant that Anna and Percy enter, early in the Scena, Donizetti's original version allowed full measures to complete the vocal lines and then full measures to open the succeeding orchestral designs (see Example 13). The resulting gaps in the continuity are especially unfortunate here, where the dramatic action, Enrico's attempt to escape the confrontation and Anna's imperious insistence that he face her, requires a dramatic and musical momentum that compel the King to remain. The revisions certainly help achieve this goal.

Within the opening section of the Terzetto, there are three small changes that also point in this direction. In the orchestral introduction, Donizetti compresses four measures of orchestral

Ex. 13. A revision in the recitative preceding the Terzetto (No. 8(b))

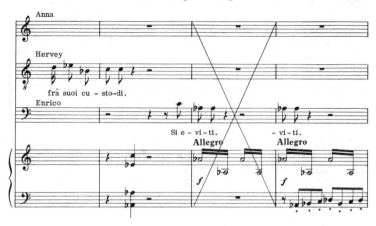

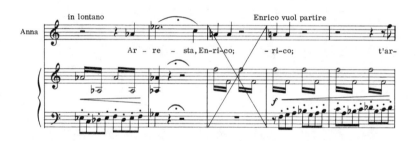

chords into two (Example 14(a)), an obvious yet appropriate alteration (see p. 48).

As in the recitative, he originally allowed a full measure of resolution after Anna's phrase 'Giustizia! è muta d'Enrico in Corte', on the tonic, and Percy's similar phrase 'E tu l'ascolta, o Re', on the dominant (Example 14(b)). Both these phrases conclude intense declamatory passages that precede a more regular orchestral melody, to which the vocal lines accommodate themselves. This is the standard Rossinian design for a small ensemble, the orchestral phrase generally coming to pervade, as it does here, both the opening section and the *tempo di mezzo* of the ensemble.[1] Realizing that these measures of resolution did

---

[1] The classical example of this procedure is found in the first-act Finale of *Tancredi* (1813) by Rossini.

Ex. 14. Three small revisions in the Terzetto (No. 8(b))

(a)

(b)

(c)

nothing but hold things up, however, Donizetti quickly eliminated them. Similarly, at the beginning of the *cantabile*, 'Fin
nell'età più tenera', the orchestral introduction originally in-
corporated three measures, the bass resolving to G in the third
measure, with Percy's entrance reserved for the following one
(see Example 14(c)). By overlapping the orchestral introduction
and the opening of the vocal line, Donizetti keeps the piece's
momentum alive. Also at the end of the *cantabile*, Donizetti
cancels an additional measure of tonic harmony in the orchestra,
though here no effort is made to overlap contiguous sections.

We emphasize Donizetti's elisions because they recur
frequently in his autograph; but revisions can also go in the
opposite direction for different, equally valid reasons. In the
orchestral introduction to the opening Coro of No. 8, Donizetti
states the theme that will subsequently be sung by the chorus.
This is a formal technique typical of Rossini choruses, to which
we shall return shortly. After two bars of a tonic chord, sustained
by brass and timpani, the strings attack the theme, originally
conceived as an eight-measure antecedent-consequent phrase (see
Example 15 on p. 50). Donizetti subsequently inserted a measure
of rest between the two parts of this theme. Considering the
extent to which the opposite process is encountered throughout
the piece, why did he do so? The answer lies in the structure and
declamation of the choral text. These are Romani's verses:

Coro I	Ebben? dinanzi ai giudici
	Quale dei rei fu tratto?
Coro II	Smeton.
Coro I	Ha forse il giovane
	Svelato alcun misfatto?
Coro II	Ancor l'esame ignorasi . . .
	Chiuso tutt'ora egli è.

In this stanza of *settenario*, *sdrucciolo* verses (those with the accent
on the antepenultimate syllable) alternate with *pieno* verses (those
with the accent on the penultimate syllable), concluding with a
tronco verse (with the accent on the final syllable). The third verse
is divided between the two parts of the chorus, the second group
of courtiers replying to the question of the first group ('Smeton'),
and the first group then seeking further information. Had

Ex. 15. Revision in the Coro preceding the Terzetto (No. 8(a))

Donizetti set the text without the extra measure of rest, the music could not have captured the sense of the dialogue. The extra measure allows Smeton's name to be heard clearly and gives the first chorus time to absorb that information before framing its next question. Donizetti realized this before he actually notated the choral parts, for he shows no hesitation there about the structure of the period.

The problem of continuity particularly engaged the composer's attention in the Duetto for Anna and Giovanna (No. 7). There is one striking example in the preceding recitative, while in the Duetto itself every passage of transition between sections involved Donizetti in serious revision. Characteristic of Donizetti's recitatives in *Anna Bolena* is the ease with which they pass from simple recitation to more musically conceived arioso or even entire lyrical periods. Though the autograph score rarely bears witness to second thoughts in these passages, one revision here does suggest the kind of process that must have been occurring constantly within the composer's mind. We have referred to this section as 'recitative', yet the word fails utterly to capture its musical and dramatic qualities. None the less, the text Felice Romani provided for Donizetti is standard recitative verse, the free alternation of *settenari* and *endecasillabi* with an occasional use of rhyme between adjacent verses. Here are the opening lines:

Anna	Dio, che mi vedi in core,
	Mi volgo a te . . . Se meritai quest'onta
	Giudica tu. (*siede e piange*)
Giov.	Piange l'afflitta . . . Ahi! come
	Ne sosterrò lo sguardo?
Anna	Ah! sì, gli affanni
	Dell'infelice Aragonese inulti
	Esser non denno, e a me terribil pena
	Il tuo rigor destina . . .
	Ma terribile è troppo . . .
Giov.	O mia regina! (*si prostra a'*
	suoi piedi)

Ignoring the poetic structure, Donizetti manipulates the verses, repeating words as necessary, in order to transform Anna's opening lines into a short lyrical period, a formal prayer. Giovanna's first words, 'Piange l'afflitta', are declaimed over the orchestral coda to the prayer. The remainder of her aside and all of Anna's following speech are set as more conventional recitative. At this point Giovanna makes her presence known to Anna. Donizetti's stage direction is much fuller than Romani's ('si appressa piangendo; si prostra a' piedi d'Anna; le bacia la mano'), and he originally intended this pantomimic action to be accompanied by a melody for strings and flute:

Ex. 16. Revision in the Recitativo preceding the Duetto (No. 7)

Only after the lyrical moment had given way again to a more traditional recitative accompaniment did Giovanna sing 'Oh mia Regina!' Returning to the passage, Donizetti crossed out this setting and added the vocal line in which Giovanna doubles the lyrical orchestral melody. By doing so, the composer achieves greater dramatic continuity, provides a more touching appeal to the Queen by Giovanna at the very outset of their confrontation, and further develops that particular mixture of recitation and arioso that defines the 'recitative' of *Anna Bolena*.

In his original version, Donizetti followed Anna's last phrase of the recitative, 'E tal facea me stessa', with an orchestral introduction of three measures, the upbeat 'Sul suo' coming at the close of the third measure:

Ex. 17. A deleted passage from the Duetto (No. 7)

⊙→ ←⊙ : Deleted bars; the upbeat, 'Sul suo', is written again in the measure preceding this cut.

This orchestral introduction set off from the recitative the beginning of the Duetto 'proper', with its more regular phrase structure arising from the poetic *ottonari* that begin precisely here. Before orchestrating the composition, Donizetti crossed out these measures and rewrote the upbeat on the preceding page of the manuscript, as part of the last measure of the recitative. This does more than eliminate an unnecessary break in the continuity. The recitative, as we have seen, time and again breaks into more melodic and rhythmically periodic phrases (Anna's prayer, Giovanna's 'Oh mia Regina', Anna's 'Ch'io compri coll'infamia la vita?', etc.). In the absence of the orchestral introduction, the 'official' opening of the Duetto passes almost unobserved, creating a continuity between recitative and closed number that is rarely achieved in Italian opera of this period.

Anna does not realize that Giovanna is her rival until the end of the first section of the Duetto. As she does so, the music moves to the dominant of B minor. Only at the last moment does this harmony dissolve into the dominant of G major, the latter being the tonality of the following *cantabile*. Finding the right musical and dramatic proportions for these transitional measures was difficult for Donizetti. Here the process involved not contraction, but rather a gradual expansion. Four levels are visible in the autograph (see Example 18 on pp. 54–56).

The first version went directly from f. 257v of the manuscript to f. 259r. Anna declaims 'Tu . . . mia rivale?' twice, first on the dominant of B minor, then modulating to the dominant of G major; the G major *cantabile* follows immediately, without the benefit of an orchestral introduction. Yet this dramatically

Ex. 18. Four versions of the transition to the *cantabile* in the Duetto (No. 7)

(a) First version

decisive moment is distinctly undervalued in the original version, while the modulation is if anything too abrupt. The second version, achieved by adding two more measures, one in the right margin of f. 257v, the other in the left margin of f. 259r, consolidates the new dominant while simultaneously allowing Giovanna a more expressive 'Ah! perdono!' to set off her confession. At this point, Donizetti perceived that the expansion of Giovanna's part had implications for the proper proportions of Anna's. Crossing out the already tormented measures on f. 257v and f. 259r (all these changes were made before the passage was orchestrated), he added a new leaf to the manuscript (f. 258), on which he notated the third version, without yet orchestrating it fully. This is particularly successful in its use of a sudden *piano*

(b) Second version

(c) Third version

(d) Fourth version

* These two measures actually went through an earlier version:

effect after the full orchestral *fortissimo* of the preceding bars, to underline Anna's astonishment and disbelief. But Anna's dramatic, unaccompanied declamation has disappeared, while the harmonic juxtaposition of the dominant of B minor and the dominant of G major is once again awkward. In his fourth version, Donizetti restores the declamation, effects a more fluent harmonic shift, and rewrites in a more expressive way Giovanna's 'Ah! perdono!' Even after all these layers his ear still sought the strongest dramatic gesture, and in place of Anna's third 'Tu?', he inserts the vastly more effective direct appeal, 'Seymour?' The added dramatic weight achieved through this series of revisions gives the moment a strength not unlike Anna's 'Giudici! Ad Anna!' before the *stretta* in the Finale 1°.

Later in the Duetto, at the end of the *cantabile*, Donizetti initially planned an orchestral coda to frame Giovanna's confession:

Ex. 19. A deleted passage from the Duetto (No. 7)

⊛→ ←⊛: Deleted bars; the note of resolution, 'è', is written again in the measure following this cut.

Obvious advantages are achieved both in dramatic continuity and effective contrast through the overlapping of Anna's 'Sorgi', the first word in the ensuing section, and the final note of Giovanna's period. For Donizetti these factors far outweigh the resulting loss of formal repose. The search for continuity here, as in all these examples, becomes a search for the right musical and dramatic gesture; only the individual moment can suggest what that gesture ought to be.

Donizetti's growing concern throughout his life with the problem of dramatic and musical continuity represents a significant shift in the underlying values of Italian opera. For Rossini it would have been unthinkable for the emotional center of an opera to fall outside the moments of expressive lyricism. For Verdi, on the other hand, it is Violetta's 'Amami, Alfredo,' which occurs in a passage entitled 'Scena' and is not part of a formal lyrical number, that crystallizes her emotions. Nothing quite so intense occurs in *Anna Bolena*, but Giovanna's 'Oh mia Regina' and much of Anna's Ultima Scena point in this direction. Likewise, the more fluid passage from one section of a formal number to another, so typical of Verdi's mature ensembles, depends on the way the composer handles the joints between these sections. Already in *Anna Bolena* Verdi could have found effective models from which to develop his own characteristic techniques. Had he consulted Donizetti's autograph, however, he would have realized the extent to which the older composer achieved this continuity through a force of will imposed on material originally conceived in a quite different manner.

VII. Problems of Balance and Proportion

A SIMILAR impetus motivates Donizetti to shorten or even omit musical periods whose original form approximates Rossinian designs. In Rossini's music symmetry and balance are crucial in defining musical proportions. Orchestral introductions, melodic periods, cadential phrases, crescendos—each has its natural place in the musical scheme and its natural relation to surrounding elements. This is why internal cuts in a Rossini composition are so destructive. Many of Donizetti's changes in the autograph of *Anna Bolena* reveal his tendency to give proportion and balance lower priority than directness of utterance, avoidance of formal repetition, and, as we have just seen, dramatic continuity. Though first conceiving periods with the proportions of Rossinian models, Donizetti can then ruthlessly destroy their classical balance. The results are not always satisfactory, but these revisions suggest directions Donizetti will later follow more successfully, and they help thereby to establish theoretical distinctions between his music and Rossini's.

An almost trivial example can demonstrate the principle in operation. A standard cabaletta consists of a cabaletta theme, a short passage preparing the repetition of the theme, the repetition itself, and vocal cadences, followed by instrumental cadences. Though in these broad terms cabalettas remain remarkably persistent from Rossini through Verdi, each composer realizes the design differently, whether melodically, harmonically, or formally. Let us examine the *non-thematic* sections, that is, the middle preparation and final vocal cadences, from the cabaletta to Smeton's solo in the Finale 1° (No. 5(a)).[1] Donizetti made cuts in

---

[1] Deciding what to call this piece is a problem. The poetry set by Donizetti includes only a single eight-verse strophe of *ottonari* (with each even verse a *tronco*), beginning 'Ah! parea che per incanto'. All the preceding text is in free verse, mixing *settenari* and *endecasillabi*. This seems to imply a recitative followed

both places while composing this aria. Here is the omitted music:

Ex. 20. Two deleted passages from the cabaletta of Smeton's [Cavatina] (No. 5(a))

(a)

by only a single musical section. Donizetti apparently wanted to provide Smeton with a *cantabile* section preceding the strophe, and so utilized the text:

> Un bacio ancora, un bacio,
> Adorate sembianze . . . Addio, beltade
> Che sul mio cor posavi,
> E col mio core palpitar sembravi,

to serve for an abbreviated *cantabile*. He also brings this text back in the section between the two statements of the cabaletta theme, more clearly affirming its earlier function as a true, though short, *cantabile*.

(b)

⊕→ ←⊕ : Deleted bars.

The first cut passage is yet another repetition of the orchestral phrase, with overlaid voice, played twice immediately before. These additional measures complete a restatement of the poetry of the *cantabile*, but the omission is not jarring, since the poetic content of the earlier lines is independent of this final one:

> Addio, beltade
> Che sul mio cor posavi,
> E col mio core palpitar sembravi.

The second cut passage is a sixteen-measure (essentially 2×8) cadential phrase which directly followed the cabaletta repeat. Donizetti crossed out both passages sometime before completing orchestration of the score. Why did he include them in the first place? And why did he then omit them? An answer to the first question can be sought in Table II:

Table II. The cabalettas of *Semiramide* (1823): arias and ensembles

Piece	Preparation for the cabaletta repeat	Cadences
Cavatina (Arsace)	3×4 + continuation on I I	$2 \times 2 + 3 \times 1$
Duetto (Arsace, Assur)	2×4 + continuation on V of vi; V→I I→V of vi	$2 \times 7 + 2 \times 2 + 3 \times 1$
Aria (Idreno)	3×4 + continuation on I I	$2 \times 6 + 2 \times 2 + 3 \times 1$
Cavatina (Semiramide)	3×4 I	$2 \times 3 + 2 \times 2 + 3 \times 1$
Duetto (Sem., Arsace)	3×4 + short continuation on I I	$2 \times 6 + 2 \times 2 + 3 \times 1$
Duetto (Sem., Assur)	Modulatory section	$2 \times 8 + 2 \times 2 + 3 \times 1$
Aria (Arsace)	2×8 (soloist and chorus) I	$2 \times 8 + 2 \times 4 + 3 \times 2$
Aria (Idreno)	3×4 + short continuation on I I	$2 \times 8 + 2 \times 2 + 3 \times 1$
Duetto (Sem., Arsace)	Different structure	$2 \times 12 + 2 \times 4 + 3 \times 2$
Aria (Assur)	2×8 (soloist and chorus) + continuation I on I	$2 \times 8 + 2 \times 2 + 3 \times 1$

Semiramide, Rossini's last opera written for an Italian theater, is a rather special work, not fully exemplifying his mature Italian style. The Neapolitan operas he wrote in preceding years (*Ermione*, *La donna del lago*, *Maometto II*) are in many respects more original, and certainly freer in their treatment of musical form. Precisely because of its retrospective and idealizing tendencies, *Semiramide* remains the best model for defining the structural conventions later generations associated with the Rossini operas. In nine of the ten numbers listed in Table II, the final cadential section of the cabaletta has three elements: it begins with a longer phrase, moves to a shorter one, and concludes with the shortest. This procedure builds a gradual cadential bridge between the thematic statement and the rapid-fire concluding cadences. As for the middle section, in those pieces where Rossini uses a non-modulating, four-measure orchestral theme over which voices declaim the words, he invariably repeats the theme three times. The middle section can differ in content, though, and when it is based on a longer (in these cases an eight-measure) phrase, the phrase may appear only twice.

Donizetti's original draft of both passages is thoroughly Rossinian. The preparation for the cabaletta repeat was:

$$\frac{3 \times 4}{I} + \text{continuation via V of vi to V of I};$$

the cadences were:

$$2 \times 8 + 2 \times 2 + 2 \times 2 + 3,$$

where the expansion in the last cadence (rather than a contraction) is typical of the post-Rossini generation. Donizetti's mind instinctively harkened back to these conventions, but his critical faculty rebelled against them. In the first case, nothing is lost. In the second, however, the revision is less satisfactory: it gives all the cadential phrases a uniformly short length and eliminates the lovely recall of thematic material from the middle section for the 2×8 cadence.

This admittedly small-scale example is indicative of how Donizetti begins with an idea closer to Rossinian terms and uses judicious (or even injudicious) cutting to alter its proportions and produce a more continuous musical and dramatic surface. Cuts of this kind are found throughout the autograph of *Anna Bolena*. In Rossini's music, as Table II demonstrates, each cadential phrase is typically played twice. Though there are exceptions, to be sure, and even occasions when Rossini writes an explicit repeat, then cancels it, the procedure is an integral part of his sense of musical architecture. It simultaneously offers the singer yet another opportunity to vary the melodic line at the repetition. Frequently composers did not bother to copy over these repeated phrases, contenting themselves with placing them within repeat signs.[2] In

---

[2] These passages can pose tricky editorial problems, for composers did not always remember to provide alternative resolutions that may be required by the logic of the individual musical lines.

his urge to press the action forward, Donizetti on several occasions crosses out repeat signs for cadential phrases whose repetition had been integral to his original plan. This happens twice in Anna's Ultima Scena (No. 11), once in 'Al dolce guidami', her first *cantabile* section, and once in the concluding cadences of the cabaletta. The main period of 'Al dolce guidami', during which Anna sings through her entire text (with a few short internal repetitions of words) can be diagrammed as in Table III(a):

Table III. Omitted cadential repeats in the Ultima Scena (No. 11)

(a) Structure of the vocal period of the *cantabile*, 'Al dolce guidami'

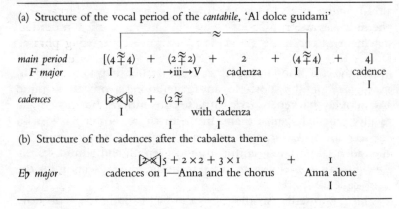

(b) Structure of the cadences after the cabaletta theme

The well-proportioned main period is immediately followed by vocal cadences that set selected verses from the full text. As is apparent in the Table, Donizetti, calling on musical proportions he had absorbed from Rossini's style, originally planned a repeat of the eight-measure cadential phrase. He sacrificed this repeat to a new vision of musical continuity, one anticipated also in his treatment of the final cadence, where the second part of what might simply have been a 2×2 cadence is varied and expanded in a way that almost disguises the underlying structural basis of the music. An absolutely parallel situation occurs in the final cadences of the Ultima Scena, after the complete repetition of Anna's solo cabaletta theme. The original structure of these cadences, through Anna's collapse, is diagrammed in Table III(b). Once again we find a series of repeated cadences of decreasing length, in the

Rossinian mold. Donizetti then cancels the repetition of the longest one, the initial five-measure cadence.

The decision to omit an exact repetition of a cadential phrase by eliminating repeat signs is, in both a material and psychological sense, one that requires minimal effort. More difficult, perhaps, is the act of canceling a varied cadential repetition, completely written-out and orchestrated. That is what Donizetti did in Anna's *cantabile*, 'Come, innocente giovane', within the Introduzione (No. 1). Unlike the previous examples, however, it is a repetition whose absence we barely perceive because of the nature of the period in which it is embedded. Anna's *cantabile* begins with an eight-measure antecedent-consequent phrase, already cited as Example 11. Instead of constructing from this phrase a symmetrical period, by concluding with a modified statement of the initial thematic idea after a contrasting phrase, Donizetti continues with a 'holding pattern' (2 × 2 measures, each moving from V of I to I) and an independent cadential phrase of six measures. In the final version, the period is followed by a cadential section of 5 + 5 measures, the second part of which is a modified version of the first, with provision for an improvised cadenza at the end. One measure of tonic harmony in the orchestra concludes the section. Originally, however, Donizetti had composed nine additional measures for the *cantabile*, a varied repetition of the 'holding pattern' and cadences in the main period. This passage and the canceled measures are given in Example 21 (see pp. 66–67).

The repetition is more one of function than of actual pitches. Indeed only in the very last measure is the vocal line identical, though the accompaniment is altered. The essential parallelism of the two passages, however, is unmistakable. Donizetti's cut actually clarifies the structure of the *cantabile*. With these extra measures, what we hear first as the conclusion of a thematic period we must reinterpret mentally as a separate cadential phrase, one that can appropriately be repeated and varied. If we add to this problem the fact that the final cadence of the *cantabile* involves a single five-measure phrase and its varied repeat, we can understand why Donizetti decided to eliminate these superfluous and ambiguous measures. The only unfortunate aspect of the cut is that as a result we never hear in the piece an orchestral figure with dotted sixteenth and thirty-second notes. When a similar

Ex. 21. Omitted varied reprise in Anna's *cantabile* in the Introduzione
(No. 1)

⊕→ ←⊕ : Deleted bars.

figure subsequently appears in the orchestra at the end of the
cantabile (| ♪ 𝄾 ♩.𝅘𝅥𝅮𝅘𝅥𝅮.𝅘𝅥𝅮 ♪ 𝄾 ♩.𝅘𝅥𝅮𝅘𝅥𝅮.𝅘𝅥𝅮 | ♪ 𝄾 𝄾 – ‖), therefore, it
seems unmotivated and inconsequential.

   Similar problems occur in the Aria for Giovanna (No. 9). Here,
too, the final cadences, after the reprise of the cabaletta theme,
were originally conceived with the proportions $2 \times 9 + 2 \times 2 +$
$(2 \times 1 + 4)$, the Rossinian design with an expansion at the very
end. The nine-measure phrase is enclosed in repeat signs, and
Donizetti simply crosses them out. More interesting is the
situation in the *cantabile*, where the composer made two significant cuts of material he had already orchestrated. First, he prepared
an orchestral introduction to the *cantabile* (Example 22), anticipat-

Ex. 22. Deleted orchestral introduction to the *cantabile* of Giovanna's
Aria (No. 9)

⊛ These two bars were cut before the passage was orchestrated.

ing the vocal theme, that he later eliminated.[3] He left untouched only the first measure.

There are countless examples throughout the operas of Rossini and Donizetti both of lyrical sections with extensive orchestral introductions and of those in which the voice enters either immediately or after a single measure of an accompanimental figuration. Interpreting this cut, then, we must consider the specific dramatic situation. Donizetti presumably did not want Giovanna, who has just admitted she still loves Enrico, to stand helpless, waiting for an extended orchestral introduction to conclude, before beginning her plea for Anna's life.

The second cut in Giovanna's *cantabile* reflects a more complex situation. Because the musical relationships involved are not simple repetitions, it is essential to show the entire period, including the measures later eliminated (see Example 23). Several measures of concluding cadences (2 + 4) have been omitted from the example.

Ex. 23. *Cantabile* period of Giovanna's Aria (No. 9)

[3] In the first six measures, the accompanimental figure was originally written for harp. Donizetti crossed the part out and rewrote it for strings, assigning the arpeggios to the second violins. From the seventh measure onward, there is no trace of harp accompaniment.

⊕→ ←⊕ : Deleted bars.

As shown in Table IV, this period is constructed as a kind of arch:

Table IV. The original structure of the *cantabile* period in Giovanna's Aria (No. 9)

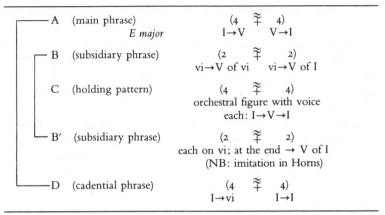

Yet there are real aesthetic problems here. Sections A, B, and D are effectively dominated by the same rhythmic figure, which can be represented abstractly as: |♩ ♪♪♪♪♪|♩ ♩ ♪ |. The slight variations along the way do not conceal its omnipresence, nor do they disguise the melodic similarities between the sections. Little wonder that the composer rebelled at his own earlier decision to return to this same material, stressing yet again the relative minor, after section C. Donizetti's form was simply too extended for the potentiality of its content. He therefore omitted B' altogether, despite the lovely imitation there between the voice and the solo horns, as well as the first four measures of D, those which made the clearest reference back to A. The resulting form of the *cantabile* is curiously difficult to grasp. One of the main problems is the deceptive cadence after the 'holding pattern', section C: did Donizetti *really* intend this effect? As an intermediary cadence closing the first half of the complete D, it is a reasonable musical gesture; as the conclusion of section C it is peculiar indeed. Whether a prospective editor of *Anna Bolena* should presume to correct the resolution is not a problem that will be discussed here. There is no doubt, however, that Donizetti's manipulations of Giovanna's period, while answering some of the aesthetic objections that could be raised to the original form, do not achieve a totally satisfactory result.

Three short examples of cuts in orchestral introductions, both to choruses and to formal musical periods, can prepare us for a more substantial example, in which similar omissions are so extensive as to threaten the internal structure of an entire composition. In the first of these examples, the cut in the orchestral introduction has a minimal effect upon the remainder of the piece; in the second, it occasions a similar cut in the vocal section; in the third, it leaves a discrepancy between the introduction and the vocal section.

The choral movement that initiates the Ultima Scena begins with an extensive orchestral introduction. After several measures of orchestral chords, a four-measure phrase is intoned by the clarinets. As he first laid out the score, Donizetti intended to have them repeat the phrase with slight variants:

Ex. 24. An omitted phrase in the orchestral introduction of the Coro (No. 11(a))

⊕→   ←⊕ : Deleted bars.

Because the orchestral introduction does not develop this material into a symmetrical period, dissolving instead into violent gestures on diminished seventh chords before reverting to a more regular cadence, the omission of the repetition of the initial phrase does not interfere with the shape of the passage itself. Nor do we feel its exclusion after the chorus enters. Although they do indeed sing the opening four measures of the orchestral introduction and then a four-measure variant of it (*not* the variant previously canceled by the composer), the music continues in a very different vein. Nothing leads us to miss the omitted measures, then, either as the orchestral introduction is first played or in retrospect.

The Introduzione of Act II (No. 6) is essentially a movement for women's chorus, consisting of an orchestral introduction, a choral period, a contrasting section, a reprise of the choral period with new text, and brief cadences. The musical structure parallels the drama: the chorus at first laments that all Anna's friends,

Table V. The structure of the woman's chorus, Act II Introduzione (No. 6)

		A				B	
orchestral introduction	7	$(4 \widetilde{\mp} 4)$	+	4	+	2×2 +	2
G major:	→I	I→I I→V		V→I		I	on I
	introductory chords	melodic phrase		cadential phrase		cadence over tonic pedal	
		A					
choral period	I	$(4 \widetilde{\mp} 4)$	+	4	+	$(4 \widetilde{\mp} 4)$	
	orch. on I	I→I I→I		on V		I→I I→I	
		melodic phrase		continua- tion		cadences	
contrasting section			26 mm.				
		A					
reprise of choral period		$[(4 \widetilde{\mp} 4) + 4 + (4 \widetilde{\Upsilon} 4)]$					
		identical to original choral period					
				B			
choral cadences		2×2	+	2×2	+	2	
		I		I		on I	
		cadence		cadence over tonic pedal		chorus	

including Seymour, have abandoned her; in the contrasting section, Anna appears, and the chorus remarks how sad and afflicted she seems; in the reprise they try to encourage her to be optimistic. The parts of this Introduzione relevant to the discussion are diagrammed in Table V.

The main thematic period of the orchestral introduction is, in the final version, twelve measures long: a four-measure phrase in the tonic, that same phrase modified so that it concludes on the dominant, and a four-measure continuation returning to the tonic. Although we might expect a further cadential phrase to follow, as indeed it did in the original version (see Example 25(a)), the proportions of the final version are perfectly acceptable:

Ex. 25. Deleted cadential figures in the Introduzione, Act II (No. 6)

(a)

⊕→ ←⊕ : Deleted bars.

⊙→ ←⊙ : Deleted bars; the two bars crossed out were eliminated by Donizetti at an earlier moment. Notice that, although Donizetti provided a resolution of the choral parts in the last measure to reflect the two-bar cut, he did not provide a resolution reflecting the longer deletion.

The main choral period, though adopting the opening phrase of the orchestral period, develops it in a very different fashion. Still, while its cadential phrase is not by any means the same as the four-measure cadence of the orchestral period, the general harmonic structure, similar orchestration, and use of parallel thirds in both passages make them seem intimately related. Thus, we have no reason to suspect anything has been cut from the orchestral introduction.

After the choral reprise, basically identical to the exposition, a cadence of 2 × 2 measures ends the vocal section; two measures in the orchestra on the tonic bring the entire piece to a close. Donizetti originally wrote a further cadential phrase at the very end, however, drawn from the omitted cadence in the orchestral introduction (see Example 25(b)). He surely decided to omit both these cadences at the same time. Structurally, the cadence at the very end of the chorus does in fact have a precise function: it helps achieve harmonic stasis more gradually, leading as it does from the harmonically active cadence that ultimately remains (2 × 2) through the cadential phrase over a tonic pedal (2 × 2), to the simple two-measure choral prolongation of the tonic. Although we can recognize the function of the omitted bars, their elimination does not seriously upset the structure of the piece.

More equivocal is the cut Donizetti makes in the orchestral material preceding Anna's prayer at the start of her scene with Giovanna (No. 7). This is a peculiar lyrical moment, as we have already seen, one imposed by the composer on lines of text not inherently meant to be set as a symmetrical period. He overcomes the problem by presenting first an extremely regular orchestral period, then allowing Anna's melody to float over a freer version of this music. The entire prayer is diagrammed in Table VI:

Table VI. The structure of Anna's prayer before the Duetto (No. 7)

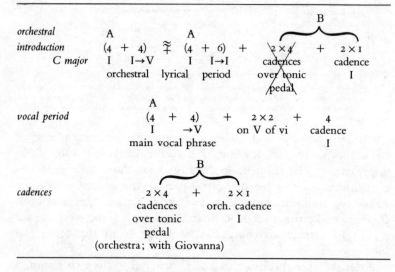

Here the cut of the 2 × 4 cadence in the orchestral introduction is much more problematic, for two reasons. First, the eighteen-measure orchestral period is insufficiently concluded by the 2 × 1 orchestral cadence that alone remains. Second, Donizetti actually uses the 2 × 4 cadence at the end of Anna's prayer. It accompanies Giovanna's entrance and her 'Piange l'afflitta . . .', to be followed by the same 2 × 1 cadence that closed the orchestral introduction. The proportions of the orchestral introduction seem rather ungainly, and the explicit use of the excised cadential theme later suggests why this should be so. Yet given the unusual nature of this lyrical moment, one without many precedents in the Rossini canon, we cannot refer to a well-formed archetypical structure as a model for Donizetti's thought. Such an archetype does exist in Rossini's music for major operatic choruses. Indeed, of the four choral movements in the 'definitive' version of *Anna Bolena*, three are typical examples of the form: the Introduzione of Act II (No. 6), the chorus that precedes the Terzetto (No. 8), and the chorus that opens the Ultima Scena (No. 11). In describing the Introduzione of Act II, we have already sketched its outlines. The small cuts the composer made in both No. 6 and No. 11 reveal his impatience with the imperatives of details of the form. This impatience erupted into rebellion in his treatment of the chorus that opens the Introduzione to Act I (No. 1). Two needs are apparent: to escape Rossinian formal requirements and to achieve greater continuity. In this number, where fundamental structural questions are involved, the limitations of Donizetti's method become evident.

The problem is posed, on a small scale, by the orchestral introduction to this chorus. Table VII is a structural diagram of the passage, as found in printed editions:

Table VII. The structure of the orchestral introduction opening the Introduzione (No. 1), as found in printed editions

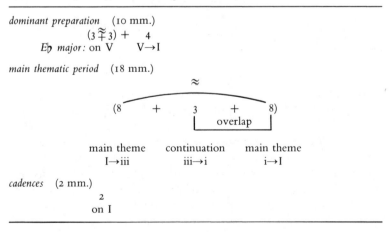

dominant preparation (10 mm.)

(3 ≈ 3) +   4

E♭ *major*: on V   V→I

main thematic period (18 mm.)

≈

(8   +   3   +   8)

|___ overlap ___|

main theme   continuation   main theme

I→iii   iii→i   i→I

cadences (2 mm.)

2

on I

A ten-measure dominant preparation precedes the main thematic period, worked out ingeniously, the tune presented first in major and repeated, after a return to the tonic, in minor, with a final cadence in major. But then, peculiarly, the music sits on the tonic for two measures and the orchestral introduction is over. There is something out of joint; the proportions seem completely misjudged. After the preparation and main period, the two concluding measures lack the requisite weight to round off the orchestral introduction. It is as if we lopped away the concluding 2×4 cadences from the orchestral introduction to the 'Introduction' of *Guillaume Tell*, leaving only a few measures of G major chords preceding 'Quel jour serein'. Donizetti had, of course, planned a cadential phrase, given in Example 26, very much like the Rossinian one just invoked:

Ex. 26. Deleted cadential phrase from the orchestral introduction of the Introduzione (No. 1)

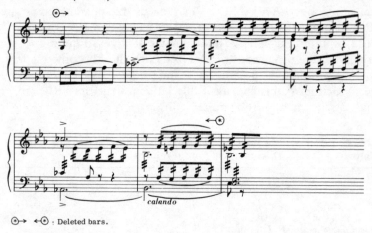

⊕→   ←⊕ : Deleted bars.

Complete string parts are present, which implies that the composer was orchestrating when he decided to omit the cadences. He had not finished the orchestration, however, since the preceding measure's wind parts remain unresolved. Even had he conceived the cadential phrase for strings alone, it is unlikely he would have neglected this resolution. The phrase is a formal closing gesture, whose loss is not felt thematically. By giving it up, though, Donizetti repudiates the need for formal balance in a context where formal balance still seems necessary. Only when the context changes, as it will in the Introduzione of *Parisina*, *Lucrezia Borgia*, or *Lucia di Lammermoor*, does the cadential gesture truly become expendable. The omission in *Anna Bolena* is indicative of a direction desired, but is not yet a successful realization of that desire.

This was only the beginning of Donizetti's dissatisfaction with his initial conception of the choral movement opening the Introduzione. The original libretto of *Anna Bolena* has a much longer text for this passage:[4]

Coro 1.	Né venne il Re?	
Coro 2.	Silenzio.	
	Ancor non venne.	
1.	Ed ella?	
2.	Ne geme in cor, ma simula.	
Tutti.	Tramonta omai sua stella.	
	D'Enrico il cor volubile	
	Arde d'un altro amor.	
1.	Tutto lo dice.	
2.	Il torbido	
	Aspetto del Sovrano ...	
1.	Il parlar tronco ...	
2.	Il subito	
	Irne da lei lontano ...	Verses absent in
Tutti.	Un acquetarsi insolito	the standard
	Del suo geloso umor.	musical text
Insieme.	Oh! come ratto il folgore	
	Sul capo suo discese!	
	Come giustizia vendica	
	L'espulsa Aragonese!	
	Fors'è serbata, ahi misera!	
	Ad onta e duol maggior.	

[4] A copy of the original libretto (ANNA BOLENA / TRAGEDIA LIRICA IN DUE ATTI / RAPPRESENTATA / NEL TEATRO CARCANO / IL CARNEVALE 1830–31 / MILANO / PER ANTONIO FONTANA / M.DCCC.XXX) is found in the Biblioteca di Santa Cecilia, Rome, Carvelhaes collection No. 965. This additional text is also found, enclosed in *virgoletti* (which implies it was omitted), in the libretto printed in the Ricordi piano-vocal score, Pl. No. 45415. My copy has the blind-stamp '6/1880'.

In the autograph, all this text was originally set and fully orchestrated in forty-six additional measures, following the E♭ major cadence at 'Arde d'un altro amor' and preceding the contrapuntal cadential phrase, 'Fors'è serbata, ahi misera!' It is unclear who indicated this cut in the autograph. Before these bars is a heavy black line, with the word 'basta' and a special sign signifying a cut; after them is a sign matching the other, with the word 'quì'. Though these words are not in Donizetti's hand, the cut was presumably approved by him. The first edition, in all its known states, already lacks these measures.[5]

What was the music like, why was this cut made, and what is the effect of the truncation? Here is the omitted music, in a reduction for piano and voices:

Ex. 27. Deleted music from the initial choral movement of the Introduzione (No. 1)

(a) The orchestral conclusion of the main choral period, the middle section, and the reprise of the main choral period

[5] We shall examine below some of the problems surrounding states of the first, oblong Ricordi edition of *Anna Bolena*.

(b) Deletion in the final orchestral cadences

⊛→ ←⊛ : Deleted bars.

The original chorus was obviously more expansive than the revision. Table VIII (see p. 88) diagrams its structure, with subsequent cuts marked.[6]

Labeling (D) a 'reprise' of the main choral period is somewhat misleading. First of all, Donizetti uses a sonata-like procédure, taking the first statement of the main choral period from I to V and the reprise from I to I (as in the orchestral introduction).

[6] There are signs of still earlier changes in the manuscript of this passage as well. The first three measures on f. 25r have been crossed out. They are similar, but not identical, to the transitional orchestral measures that lead back to the main theme in Example 27. A different leaf from the present f. 24 must once have preceded them. The present f. 24, on which the music is crowded together as if Donizetti knew he had to squeeze a certain number of measures in a predetermined space, connects directly to the bar of f. 25r following the deleted bars.

Table VIII. Original and modified structure of the chorus opening the Introduzione (No. 1)

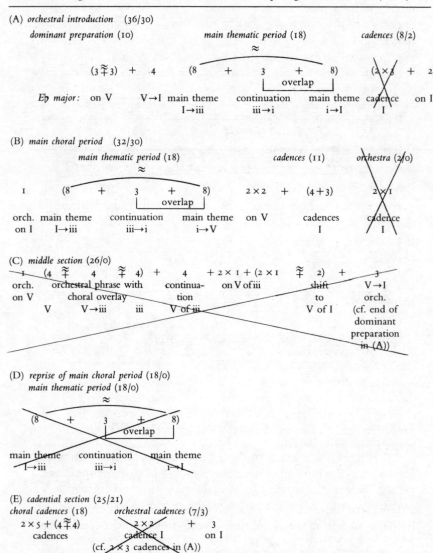

(A) *orchestral introduction*   (36/30)

    dominant preparation (10)   *main thematic period* (18)   *cadences* (8/2)

                                   ≈

        (3 ≋ 3) + 4   (8   +   3   +   8)   (2 × 3 + 2)
                                   overlap

E♭ *major:* on V   V→I main theme continuation main theme cadence on I
                       I→iii   iii→i   i→I   I

(B) *main choral period*   (32/30)

           main thematic period (18)   *cadences* (11)   *orchestra* (2/0)

                      ≈

I   (8   +   3   +   8)   2 × 2 + (4 + 3)   2 × 1
                     overlap

orch. main theme continuation main theme on V   cadences   cadence
on I   I→iii   iii→i   i→V   I   I

(C) *middle section* (26/0)

   4   (4 ≋ 4 ≋ 4) + 4 + 2 × 1 + (2 × 1 ≋ 2) + 3
orch. orchestral phrase with continua- on V of iii   shift   V→I
on V   choral overlay   tion   to   orch.
    V   V→iii   iii   V of iii   V of I (cf. end of
                                           dominant
                                           preparation
                                           in (A))

(D) *reprise of main choral period* (18/0)

    main thematic period (18/0)

              ≈

(8   +   3   +   8)
           overlap

main theme continuation main theme
I→iii   iii→i   i→I

(E) *cadential section* (25/21)

choral cadences (18)   *orchestral cadences* (7/3)
  2 × 5 + (4 ≋ 4)   2 × 2   +   3
    cadences   cadence I   on I
         (cf. 2 × 3 cadences in (A))

Furthermore, the choral part is entirely different. Whereas in the main choral period the chorus sang whispers of court rumor under its breath, with the darkness of the words matched to the furtiveness of the musical lines, in the reprise the chorus sings out the melody, emphasizing the contrast between the major and minor modes and the triads built on their respective third degrees (G minor concluding the first phrase, G♭ major so prominently displayed in the second).

The periods differ in another way. With its wind and string dialogue over accompanimental rhythmic figures, the middle section, while effective in itself, does not sit alone. Its orchestration informs and transforms the reprise of the main theme, for the accompanimental rhythm (| ♫ ♪ |) is added throughout, played first by strings and timpani, then, for the statement in the minor, by timpani and cello alone. Timpani and cello continue the rhythm during the (4 + 4) cadence and during the short (2 × 2) orchestral cadence, ultimately omitted (Example 27(b)), which hearkens back to the one Donizetti cut at the close of the orchestral introduction (Example 26).[7] The orchestral accompaniment of the (4 + 4) cadence, in fact, loses much of its point without the middle section that first introduces this rhythm.

Donizetti's original conception of the chorus was thoroughly in the Rossinian mold in its outward appearance. This is, in fact, Rossini's standard way to construct any chorus, especially one opening an Introduzione or Finale, and similar examples abound throughout his operas.[8] It should come as no surprise to find Donizetti instinctively drawn towards this formula. But his particular realization of the Rossinian model here is extremely inventive and unusually successful in making the form function not only architecturally, as in Rossini, but dramatically, producing continuity and change within the chorus by manipulating the very limitations of the genre.

Donizetti ultimately opted against this fuller choral movement. Perhaps he felt the male chorus at his disposition incapable of

[7] The impetus behind these cadential cuts is similar to that behind the cadential cuts made in the Introduzione of Act II (No. 6). See Example 25 and the related text.

[8] I have discussed some of the characteristics of the archetypical Rossinian Introduzione in my article cited above, 'Gioachino Rossini and the Conventions of Composition'.

sustaining such an extended piece. The three choruses of Act II are all much shorter and simpler.[9] Perhaps he felt the number, despite its inventiveness, was too long for the dramatic situation and too reminiscent of Rossini. Perhaps he simply felt that the music was poor, though that conclusion is difficult to accept. In any event, he cut it drastically, removing both the middle section (C) and choral reprise (D), which thus, late in the history of the piece, join the cadential cuts in the orchestral introduction and final cadences, made before the orchestration was finished. These cuts all have a similar result: to speed up the choral opening and get on with the business of the opera. But they also have a rather unfortunate effect on its proportions. With the orchestral introduction (A) and cadences (E) essentially intact, Donizetti is left with a main choral period (B) unable to hold its weight against the remainder of the composition, particularly since the choral parts are, for dramatically sound reasons, barely formed. The cadential section (E) carries all the structural weight of typical Rossinian cadences, but serves as ballast for a much shorter movement. What is left resembles its model too closely, so that while the cuts can alter the proportions, they cannot make us forget what those proportions ought to be.[10]

When a composer actively seeks to go beyond musical procedures that seemed congenial to an earlier generation but were growing increasingly restrictive to his own, he inevitably passes through a period of experimentation. The situation is particularly difficult when the earlier procedures embody a perfection of formal control such as that found in the music of a Rossini or a Mozart. There is something almost willful in many of the compositional decisions Donizetti takes in these examples. He could not have been unaware of the damage he was doing to the musical proportions he had so meticulously projected. Though a new sensibility is apparent here, it does not yet know how to define new formal procedures adequate to its needs. In some cases,

[9] Though it is clear that the Teatro Carcano in this extraordinary season attracted the finest soloists of the time, we do not know whether orchestra and chorus were of equal merit.

[10] Near the end of this study, we shall return to a consideration of the structure of the entire Introduzione.

Donizetti's cuts do succeed in moving the drama along at a faster pace without adversely affecting his music; in others, the residue of a once noble conception can seem almost incomprehensible. There are few works of art that do not suffer from such tensions. To ignore them or to explain them away falsifies and trivializes the work of art and the process by which it was created. Writing music is, and always has been, hard work.

VIII. Tonal Closure

TONAL coherence within individual numbers was a matter of urgent concern to Rossini. This usually means that, in his operas, each individual musical number begins and ends in the same key, whether the number is a simple aria, the *cantabile* and cabaletta jutting directly up against one another, or a massive ensemble in many distinct parts, such as the great 'Terzettone' of *Maometto II*[1] or the Terzetto of *La donna del lago*. Only occasionally in Rossini's operas is the principle weakened. Arias preceded by introductory choruses, for example, are found with two different kinds of tonal schemes. In some cases, the chorus and concluding cabaletta are in the same key, while the *cantabile* solo movement between them is in a related tonality; in others, the aria itself begins and ends in the same key, while the introductory chorus is in a related tonality. An example of the latter, much the most typical arrangement in Rossini's operas, is Isabella's Coro e Cavatina in *L'italiana in Algeri*, where the introductory chorus is in C major, but both the *cantabile* and cabaletta of Isabella's Cavatina are in F major.

The gradual dissolution of this constructive principle played an important role historically in weakening the boundaries between distinct musical numbers in Italian opera of the nineteenth century. *Anna Bolena* is still Rossinian enough to adopt strict tonal boundaries for most numbers. Only two numbers in the opera lack tonal closure: Percy's Cavatina (No. 3) and the Ultima Scena (No. 11). Other apparent exceptions occur in different situations. The first-act Finale, for example, is effectively constructed of three separate numbers, a Cavatina, Duetto, and the Finale 'proper', each of which is tonally closed (A♭ major, B♭ major, and D major, respectively). The Coro preceding the Scena and

---

[1] I discuss this composition briefly in the Introduction to the facsimile edition of *Maometto II* (New York, 1981), in the series Early Romantic Opera, ed. Philip Gossett and Charles Rosen.

Terzetto (No. 8) is in E♭ major, while the Terzetto itself is a closed form in C major. It is similar, therefore, to the problem of the Rossinian Coro e Cavatina mentioned above. The Ultima Scena cannot be explained away by such means. After an opening chorus in F minor, Anna's recitative begins with an orchestral period in F major, also the key of her first *cantabile* movement, 'Al dolce guidami'. The march accompanying the arrival of Percy and Rochefort is in A♭ major, while Anna's next *cantabile*, 'Cielo, a' miei lunghi spasimi', is in G major. Though A♭ major recurs as the key for the off-stage music celebrating the marriage of Enrico and Giovanna, the cabaletta of the Ultima Scena, 'Coppia iniqua', closes the number in E♭ major. One might well argue that this choice of key at the conclusion of the opera depends on tonal considerations overriding the final number: that is, a return to the tonality of the Act I Introduzione, E♭ major. But the function of large-scale tonality in the operas of Donizetti remains for now *terra incognita*.

That the two principal parts of Percy's Cavatina should be in different keys is even more striking, because the composition, with its *cantabile*, *tempo di mezzo*, and cabaletta, is so clearly linked to Rossinian traditions. The role of Percy was created by Giovanni Battista Rubini, whose upper register was legendary. It should not therefore surprise us to find both his solo numbers transposed downwards in printed editions of *Anna Bolena*, the Cavatina (No. 3) from B♭ minor/E♭ major to G minor/C major, the Aria (No. 10) from A major to G major. In Bellini's Rubini roles, too, which included the title part in *Il pirata*, Elvino in *La sonnambula*, and Lord Arturo in *I puritani*, we find similar downwards transpositions in printed scores. Some of these go so far as to subject different sections of the same aria to different transpositions. Elvino's second-act aria in *La sonnambula*, to take but one example, is printed in modern editions with its *cantabile* in A minor/C major and its cabaletta in B♭ major. In Bellini's autograph, the aria begins in B minor/D major and concludes in D major, preserving that tonal coherence the modern transposition discards. The high Ds in the original key of the cabaletta are, to be sure, enough to frighten away modern tenors, but it seems clear that the entire composition should be transposed down by the same interval. In fact, the first edition, published by Ricordi in 1831, brought the keys down to B♭ minor/D♭ major and D♮ major, certainly a preferable alternative to the confusing modern key scheme.

There is a strong possibility, however, that Percy's Cavatina was originally intended to be a compositon with tonal closure. Irregularities in the manuscript suggest this. Problems first surface in the *tempo di mezzo*. Through this point the Cavatina is written on a single kind of paper, grouped in two gatherings of two bifolios each (ff. 90–3 and 94–7 of the autograph). Folio 98 is of the same paper type. At the end of its recto, Donizetti has crossed out two measures of music; he has also crossed out all the music on its fully notated verso. This extra music, unquestionably the original continuation of the *tempo di mezzo*, given in Example 28(a), is most striking for its heady insistence on F major as the dominant:

Ex. 28. Two passages from the autograph of Percy's Cavatina (No. 3)

(a)

⊙→ ←⊙ : Deleted bars.

With the main material of the *tempo di mezzo*, the hunt, in the subdominant, E♭ major, a return to the original B♭ major for the cabaletta would not only make sense but would also, through giving the piece tonal closure, follow the older tradition.

Folio 98 is an isolated leaf in the autograph; it is followed by yet another isolated leaf, f. 99, on different paper. We therefore cannot know how the music of f. 98 originally continued, although we can be certain the cancellation of the bars on f. 98 and the substitution of the successive folios occurred late in the history of the opera, perhaps even after the first performances, since the material crossed out on f. 98 is completely orchestrated. The verso of f. 99 picks up after the measures not crossed out on f. 98r, completing in Donizetti's hand the music preceding the cabaletta. On the recto of f. 99, however, we find the orchestral conclusion of a composition in B♭ major, as in Example 28(b). For obvious reasons it would be appealing to consider this passage the orchestral conclusion of an earlier cabaletta for Percy. Though there are two vocal staves between the cello and bass parts at the bottom of the score and the timpani staff above, the important role Rochefort plays in the Cavatina would easily justify his having sung during the final cadences of this presumptive cabaletta. Thus the hint of the *tempo di mezzo* conclusion of Example 28(a) would be supported, and we could presume that Donizetti originally followed more closely to Rossinian tenets, closing the piece on the tonic with which it began, B♭.

There is one particular aspect of the instrumentation, however, that might argue against this interpretation and suggest that the music found on f. 99r has nothing whatsoever to do with a possible original cabaletta to Percy's Cavatina. The orchestration and physical arrangement of the instruments in the score, a relatively standard one, is the same on f. 99r as in earlier parts of the Cavatina (ff. 90–8): violin I, violin II, viola, piccolo, flute, oboes, clarinets in B♭, horns in E♭, trumpets in B♭, bassoons, trombones, timpani, voices, cello, bass. Throughout the *cantabile* and *tempo di mezzo*, the two timpani available to Donizetti are tuned to the pitches B♭ and E♭, and Donizetti writes these explicit notes in his autograph. This is not a standard arrangement for a composition that begins in B♭, whether minor or major. Much more typical would be a tuning on the tonic and dominant of the tonal center, B♭ and F in this case. The importance of the subdominant in the hunting music of the *tempo di mezzo* might justify this tuning even had the original cabaletta been written in the key of B♭ major, since the pitch E♭ can adapt itself either to IV (as the tonic) or V (as the seventh degree—less than ideal, but possible). A timpani tuned to F, on the other hand, would have been useless for the subdominant passages. It cannot be excluded, however, that the tuning to B♭ and E♭ reflects an early compositional decision to write the cabaletta in E♭ major, despite the lack of tonal closure.

Given this context, there are two peculiar aspects of Donizetti's notation for the timpani parts in the B♭ major cadences found on f. 99r. First of all, they are tuned to the pitches B♭ and F, not to B♭ and E♭. Nowhere else in *Anna Bolena* does Donizetti change the tuning of his timpani within a single number, although when there is sufficient time he is willing to alter the tuning between an introductory chorus and the following more or less independent number. (In the Coro, Scena e Terzetto, No. 8, the timpani are in

E♮ for the Coro and C for the Terzetto; in the Ultima Scena, they are in F for the opening chorus and E♭ for the remainder of the number. When we refer to timpani being 'in a key', we mean that the two instruments are tuned to the tonic and dominant of that key.) His feeling about this matter apparently went so far as to compel him to use timpani tuned in D major for the Duetto finale, a piece in B♭ major in both its forms (No. 5(b) and No. 5(b)'), rather than require the instruments to retune from B♭ major to D major in time for the Finale 'proper' (No. 5(c)). Second, whereas earlier in the Cavatina the timpani are notated explicitly with the pitches B♭ and E♭, in these cadences on f. 99r, they are written 'C' and 'G'. This was not an unusual notation in Donizetti's era. The tuning was shown at the opening of the number, much as in the case of other transposing instruments, and it was presumed that the two timpani were tuned to the tonic ('C') and the dominant ('G'). In fact, throughout *Anna Bolena* Donizetti uses the 'C–G' notation *except* when his timpani are tuned either to E♭ major or E major, in which cases he writes the exact pitches. The history of this notational practice and the extent to which it is found throughout Donizetti's music remains unknown. Still, even granted it may have been normal to notate timpani in E♭ on the pitches E♭ and B♭ and timpani in B♭ on the pitches C and G, it seems very peculiar to encounter both forms of notation successively within the same composition.

For these reasons, the notation of the timpani parts on f. 99r casts doubt on the theory that these cadences in B♭ major were once associated with an original version of the cabaletta to Percy's Cavatina. Whether the autograph of the cabaletta would offer any further clues cannot now be determined, for only the standard, E♭ major cabaletta is found in the autograph score, on ff. 100–104r, and it is in a copyist's hand. The five-measure orchestral ritornello opening the cabaletta is missing in the copy, however, and on f. 100r Donizetti (?) noted 'Il ritornello è in fine'; on the verso of the final leaf of the copyist's score (f. 104v), the composer himself wrote out the ritornello. This demonstrates unequivocally that he approved use of this cabaletta, even if a different piece, perhaps one adhering more closely to Rossinian tonal conventions, may once have existed. A thorough examination of Donizetti's earlier operas might furthermore reveal that this standard cabaletta is in fact borrowed from another opera by

Donizetti: that would explain why it exists in a copyist's hand in the autograph of *Anna Bolena*.

The problem of tonal closure in formal numbers is absolutely fundamental to our perception of the musical structure and continuity of an opera. There are conflicting forces at work: the classical concern for order that establishes a separate key for each number, a concern that makes perfect sense in an operatic style where formal numbers are separated by secco recitative, as in Rossini's early works; and a concern for continuity and large-scale tonal reference, in which the tonal movement from one number to another, all music accompanied by the orchestra, diminishes the sense of division between successive numbers, so that keys could take on symbolic or structural significance in the work as a whole. Verdi's use of C minor and Db minor/major in *Rigoletto* is the most obvious example of such large-scale tonal planning. Though Donizetti's willingness to throw off tonal closure in individual numbers is only a first step in this direction, it is an important one. His intention in the case of Percy's Cavatina, if we have interpreted the evidence correctly, may have been nothing more than the desire to replace a cabaletta that did not appeal to the singer or the public; no matter. The opera as sung and printed proclaimed that tonal closure within an individual number was no longer sacrosanct. In later works Donizetti himself and other composers could draw further lessons from this and similar examples found in operas written during the late 1820s and early 1830s.

IX. Wrestling with the Cabaletta

THOUGH we cannot trace in detail Donizetti's revisions in the cabaletta of Percy's Cavatina, there is ample evidence in the autograph score of *Anna Bolena* that the problem of the cabaletta more generally was one of the crucial issues faced by the composer in his work on the opera. The cabaletta, for better or worse, dominated Italian opera from Rossini to Verdi,[1] though by 1830 it was already the butt of critical scorn, as we saw in contemporary reviews of *Anna Bolena* and in Donizetti's letters. This did not prevent Donizetti from writing twelve cabalettas for the standard version of *Anna Bolena*, one for each number except the Act II Introduzione (No. 6), which is purely choral, but three in Finale 1°: for the Smeton Cavatina, for the Duetto of Anna and Percy, and for Anna and the ensemble to conclude the act. He also prepared a cabaletta for the 'Duetto rifatto' (No. 5(b)'). Of these, nine are formally regular, with two full statements of the theme separated by a short middle section. Four are irregular: the cabaletta for the Duetto of Giovanna and Enrico (No. 2), whose reprise is abbreviated; that of the Quintetto (No. 4), which has no reprise; that for the Duetto of Anna and Percy (No. 5(b)'), where a rapid concluding section takes the place of the reprise; and that for the Duetto of Anna and Giovanna (No. 7), whose final section fulfills the function, but does not suggest the shape, of a cabaletta. Significantly these problematic cabalettas are present in most of the ensembles of the opera, for which Donizetti was unmistakably seeking forms he considered more dramatically compelling.

During the composition of all four of these pieces, Donizetti faced extensive problems over the cabaletta, problems which arose not from the inherent difficulties of writing cabalettas but

---

[1] Verdi's attitude towards the cabaletta and the extent to which cabalettas still influenced his style in *Aida* are among the matters discussed in my 'Verdi, Ghislanzoni, and *Aida*', op. cit.

rather from the inherent difficulties of not writing them. No significant alterations are found in the autograph for the cabaletta of the Duetto of Anna and Giovanna (No. 7), but two surviving manuscripts of the cabaletta alone provide insight into Donizetti's original plans. In all the other cabalettas with variants of structure—the Duetto for Giovanna and Enrico, the new Duetto for Anna and Percy, and the Quintetto—the autograph itself reveals a wealth of revisions. Noteworthy problems occur too in cabalettas of the Terzetto (No. 8(b)) and Finale 1° (No. 5(c)). In this chapter we shall examine four of these problematic cabalettas: Finale 1° (No. 5(c)), Quintetto (No. 4), Duetto Anna and Giovanna (No. 7), and Terzetto (No. 8(b)). The other two pieces, the Duetto for Giovanna and Enrico (No. 2) and the new Duetto for Anna and Percy (No. 5(b)′), will be taken up in later chapters.

(a) Finale 1°

Only rarely does Donizetti seek dramatic justification for the repetition of the cabaletta theme, preferring simply to let the form work itself out in musical terms. That he was aware of the problem, though, is clear from some notably successful examples, such as Edgardo's 'Tu che a Dio spiegasti l'ali' from *Lucia di Lammermoor*: the hero's suicide motivates a cabaletta reprise in which the orchestra plays the tune while Edgardo struggles with death. Donizetti also sought to prepare the repetition of 'Ah! segnata è la mia sorte' from Finale 1° of *Anna Bolena*, and he succeeded in creating a powerful dramatic justification. Unfortunately, his solution has been confused and distorted in modern editions.

In the standard Ricordi piano-vocal score, the following stage direction is given over the measures preceding Anna's reprise of her cabaletta theme: 'affannosa segue Enrico, egli la guarda bieco' ('anguished, she follows Enrico; he looks at her malevolently'). At her last 'Ah!' before the reprise, she is instructed: 'porta la voce e rinforza.' William Ashbrook, in a discussion of this passage and the remainder of the Finale, remarks:

Enrico, although still on stage, remains silent for the last 12 pages of the ensemble [that is, from the reprise of the cabaletta theme through the end of the act]. It is a powerful insight on Donizetti's part that at this moment the king's fury has reached such a pitch that it has become inexpressible in words.[2]

It might have been a powerful insight on the part of some Ricordi editor, but Donizetti's it was not. Both the autograph and the Ricordi first edition, in all its various stages, give the direction: 'Enrico la guarda bieco *e parte; allora disperata* porta la voce e rinforza' [italics mine]. Enrico's departure justifies Anna's desperate outburst and the repetition of the cabaletta theme. It is hard to imagine where the other tradition arose, but it is found in no authentic musical source. I can attest to a recent production of *Anna Bolena*, however, in which a famous singer playing the part of Enrico, although informed of Donizetti's wishes, refused to leave the stage. That Anna might be left alone to reap audience approval at the end of the act was too hard a burden for his tender ego to accept. It would not be at all surprising were similar sentiments ultimately responsible for the 'tradition' documented in the recent Ricordi editions.

At an earlier stage in the history of the opera, in fact, Donizetti intended Enrico to remain on stage until the conclusion of the act. The end of the preparatory section for the cabaletta repeat occupies f. 197 in the autograph. The repetition itself starts on f. 198r, but this folio begins with three measures, given in Example 29(a), crossed out by Donizetti. Since the canceled measures of f. 198r follow directly neither from f. 196 nor f. 197, it seems clear that one or both of these single leaves in the autograph (ff. 192–5 form a gathering of two bifolios, ff. 198–9 a single bifolio) must have been a later addition to the score, replacing an earlier leaf or leaves that connected directly with the three canceled measures of f. 198r:

Ex. 29. Two passages from an earlier version of the cabaletta in Finale 1°

(a)

[Un detto] so - lo... per pie - tà...   Ah!   se -

Vc.e Cb.

⊕→ ←⊕ : Deleted bars.

[2] See p. 18 of the program booklet prepared for the ABC Records recording (ABC/ATS 20015/4), 1973, with Beverly Sills in the title role.

(b)

This is a much less dramatically intense version of what will become the chromatic ascent leading to the reprise. Yet dramatic intensity is surely essential if the composer is even to approach the preparation for the first statement of the cabaletta theme, the famous 'Giudici! ad Anna! ad Anna! . . . Giudici!' At this stage, furthermore, there is no mention of Enrico's departure. In fact, where he had been silent throughout the statement of the cabaletta theme, Enrico was *not* to be silent during its reprise. As is typical in Donizetti autographs, only a bass line and blank measures mark off the passage to be repeated, with one exception: a part is added for Enrico, given in Example 29(b), beginning in the fifth measure after the ensemble enters and lasting until the 'Più mosso' cadences. Enrico's line was, of course, crossed out by Donizetti when the composer decided to have the King stalk off the stage, thereby justifying Anna's final, harrowing outburst.

The changes in the Finale result from Donizetti's search for musical and dramatic motivation of Anna's repetition of her cabaletta theme, but the composer shows no obvious dissatisfaction with the form. In the next two examples, he seeks to escape the tyranny of the musical form itself. In each case he does so in quite a different fashion, suggesting alternatives to the classical model; this approach would soon find important reverberations in Verdi's style.

(b) Quintetto

Donizetti conceived the Quintetto with an essentially traditional cabaletta design, even if we cannot now entirely reconstruct this original version. The music as printed in all editions is outlined in Table IX:

Table IX. Quintetto (No. 4): structure of the cabaletta

---

introduction to the cabaletta (all)   (30)

	opening phrase		*prolongation of V*		
	(8 ⁀ 8)	(8 + 3 + 2 +	2)		
					‖ overlap
C major:	I	vi	V⁷	on V⁷	V⁷	V→I
	each: all→soloists	canonic		hold	orchestra
		(all)			alone

$$I \quad vi \quad V^7 \quad \text{on } V^7 \quad V^7 \quad V{\to}I$$

main thematic period (35)

(8 ⁀ 7)	+	(4⁀4)	+	4	+	2×4
overlap‖						
I	I→V of iii	→V	→I	I		I
Percy	Anna and Percy	continuation		sequential		cadence
		a 5		cadence		*a 5*

middle section (??)   (34)

2 × 6	(6 + 2 + 6 + 2 + 6)
phrase split among	c r e s c e n d o
chorus and soloists	I
each: I→ii→V→I	

cadences   (65)

vocal cadences	*orchestral close*	
2 × 18 + 2 × 4 + 3 × 2 + (2 × 2 ⁀ 3)	2 × 2	+ 2 × 2
cadences for all on I	arpeggiation	I
	I	

By the standards of ensemble cabalettas in Rossini (that which
concludes the Introduzione of *Semiramide*, to take but one
example, has many affinities with this Quintetto), the passage
labeled 'middle section' is peculiar for two reasons: (1) it *is* so
obviously a middle section we inevitably wonder what happened
to the reprise; (2) it is really two middle sections in one, with the
initial phrase wanting to prepare a reprise, not a crescendo, which
itself wants to be the main event of a middle section. The
cadences, furthermore, seem grossly out of proportion to the
thematic period.

This is another example of Donizetti's attempting to forge a
personal formal approach using musical material resistant to the
purposes he presses it to serve. That he juggled his material about
is evident in the autograph. There are no problems in the

introduction to the cabaletta, the main thematic period, or the opening 2 × 6 of the middle section. This material continues until f. 127v of the autograph, at the end of which are four crossed-out measures, given in Example 30(a), with vocal parts, cello, and bass lines alone:

Ex. 30. Remnants of an earlier cabaletta continuation for the Quintetto (No. 4)

(a) Deleted passage continuing the middle section of the cabaletta

Folio 128r, the next folio in the manuscript as now constituted, continues with the four measures in Example 30(b), also crossed out, followed directly by the cadences. All the paper in the Quintetto is of the same type. The first part of the cabaletta is a gathering of two bifolios, the normal unit for non-problematic places in this autograph. Folios 127–8 form a single bifolio. Since

Ex. 30(a) *cont.*

⊕→ ←⊕ : Deleted bars.

the music on f. 127v clearly does not continue on f. 128r, it seems likely that here too Donizetti utilized a gathering of two bifolios, the middle bifolio of which was then removed. What did it contain?

The music on f. 128r, Example 30(b), is actually a variant, with added chorus, of the close of the cabaletta theme. (The harmony in the second measure remains centered around the dominant degree, whereas in its earlier appearance the subdominant was featured.) This is surely the ending of what was once a varied repetition of the· cabaletta theme. The music preceding the manuscript disjunction, Example 30(a), seems like a phrase that would inevitably have led to a dominant preparation for the repeat. Provided the dominant was reached quickly, there would have been ample room in the missing bifolio to accommodate this

Ex. 30(b) Deleted passage closing a varied repetition of the cabaletta theme

⊕→ ←⊕ : Deleted bars.

music. Thus, Donizetti originally composed a normal cabaletta. Rejecting this approach early in the history of the opera, perhaps as a conscious reaction against formulaic demands, he never orchestrated the passage. (There would, of course, have been no reason to orchestrate the cabaletta repetition, except to accommodate changes, but Donizetti would have had to orchestrate the music of Example 30(a).) When he decided to remove the offending section, he simply lifted out the bifolio forming the central element of the original gathering.

There is no provision in the autograph, however, for the crescendo, found in all printed editions between the 2×6 phrase and the final cadences. That Donizetti wrote it is proven by the separate autograph part for bass drum and timpani. Because there was no room for these parts in the full score, Donizetti notated them on a separate single bifolio, later used as a wrapper for the entire Quintetto, thus forming ff. 106 and 131 of the autograph. Here, on f. 131r, the twenty-two measures of the crescendo are included (the figure, with its two-measure bridge between statements of the main crescendo theme, is extremely characteristic and easy to identify, even from a percussion part). But Donizetti has crossed these measures out. If they were once actually present in the main manuscript, they too must have been located between ff. 127 and 128. We seem therefore to have the anomalous situation in which two separate bifolios were used successively as the inner bifolio of a gathering of two bifolios whose outer bifolio was the present ff. 127–8 of the autograph score: (1) the original inner bifolio, containing the cabaletta repetition in an unorchestrated state; (2) the substitute inner bifolio, containing the crescendo, fully orchestrated. When Donizetti decided also to omit the crescendo, he removed the substitute inner bifolio, as he had once removed the original one. This decision must have followed prepartion of the first edition of the piano-vocal score, from which later editions are derived.[3] One well understands Donizetti's unhappiness with this superfluous crescendo. The composer seems caught in a web of Rossinian devices, unable in this piece to do more than rearrange the elements in a not completely successful manner.

A composer seeking alternatives to the standard cabaletta structure could not simply omit the repetition of the theme, thereby weakening the proportions and balance in the composition as a whole. He needed, rather, a way of preserving the importance and function of the cabaletta while allowing its internal structure to bend to dramatic needs. This was the direction Donizetti took in the Duetto for Anna and Giovanna.

(c) Duetto Anna e Giovanna

Although he does not fully understand its formal structure, Ashbrook is certainly correct when he states that this Duetto

---

[3] The crescendo is also found in a Roman manuscript copy, at the Biblioteca di Santa Cecilia, G. Mss. 735–6.

'with its clear dramatic progression, its interrelationships heightened by motivic and rhythmic echoes, is one of the highpoints of *Anna Bolena*, revealing Donizetti's efforts to escape from formal convention toward dramatic truth.'[4] Donizetti works throughout at differentiating the characters rather than reducing their conflict to parallel melodic phrases. Examination of the autograph would lead one to believe that the composer experienced few compositional difficulties in working out the cabaletta, with its absolutely distinct melodic periods for the two women, Anna's 'Va, infelice, e teco reca' expressing her forgiveness, and Giovanna's 'Ah! peggiore è il tuo perdono' revealing her anguish and guilt. But the autograph is deceptive. Two manuscripts, both in the collection of The Pierpont Morgan Library in New York, attest to the tortured history of this particular cabaletta. Establishing a chronology for the manuscripts is difficult, for each is close in a different way to the final autograph. They are, in fact, not sketches but rather autograph copies of Anna's part prepared by Donizetti for his prima donna, Giuditta Pasta.[5]

What is presumably the earliest version, for reasons to be considered below, is a single bifolio notated on all four pages. It contains Anna's part in full, with almost all the accompanying bass line; Giovanna's part is copied hurriedly but completely when she sings alone, and very incompletely or not at all when she sings together with Anna. The second manuscript, also a single bifolio, is notated only on the first three sides. Though written more neatly, it incorporates some revisions: in particular, on two occasions Donizetti inserts additional measures by extending staff lines into the margin and squeezing additional

---

[4] Ashbrook, *Donizetti* (1982), p. 257.

[5] I would like to express my deep thanks to J. Rigbie Turner of The Pierpont Morgan Library for his assistance in providing me with access to these manuscripts. In the catalogue of *The Mary Flagler Cary Music Collection*, published by the Morgan Library in 1970, only one of these manuscripts is listed, as N. 101a in the Addenda on p. 105. A facsimile of the first page is given as Plate XX. In preparing an up-to-date listing of the musical autographs in The Morgan Library, published as 'Nineteenth-Century Autograph Music Manuscripts in The Pierpont Morgan Library: A Check List', *19th-Century Music*, iv (1980), 49–69 and 157–83, Mr. Turner discovered that there were indeed two manuscripts, which he calls, incorrectly I think, 'drafts' (p. 65). The association with Giuditta Pasta is made explicit by the cover title: 'Scena dell'opera *Anna Bolena* per Giuditta Pasta'.

music on his hand-drawn staves. The vocal lines are written more completely than in the first manuscript, and the bass is included throughout. Still there are lacunae in Giovanna's part. Near the end, Donizetti leaves several measures blank, indicating only '3' to suggest she sings a third above Anna. Even when her notes are explicit in the cadential passage, no text is provided for Giovanna. Again, the manuscript seems to have been prepared for the use of Giuditta Pasta.

The survival of these two 'particelle' for Anna may indicate that a complete 'particella' once existed from which Pasta studied her part. Each time Donizetti decided to replace the cabaletta he removed the superseded bifolio. Presumably the final version of her 'particella' corresponded to the full autograph score. Quite apart from their musical content, the Morgan manuscripts confirm again that such 'particelle' were indeed utilized by singers in learning their roles,[6] and they prove that Donizetti wrote at least some of them himself.

Romani's text, constructed to accommodate a traditional cabaletta, consists of absolutely parallel stanzas for the two characters:

Anna	*Giovanna*
Va, infelice, e teco reca	Ah! peggiore è il tuo perdono
Il perdono di Bolena . . .	Dello sdegno ch'io temea.
Nel mio duol furente e cieca	Punitor mi lasci un trono
T'imprecai terribil pena . . .	Del delitto ond'io son rea.
La tua grazia or chiedo a Dio,	Là mi attende un giusto Dio
E concessa a me sarà.	Che per me perdon non ha.
Ti rimanga in questo $\lbrace$ addio / amplesso	Ah! primiero è questo $\lbrace$ addio / amplesso
L'amor mio—la mia pietà.	Dei tormenti che mi dà.

The surviving sources, however, bear witness that Donizetti never planned parallel musical settings for them.

In the earliest version, the cabaletta is in C major throughout, giving tonal closure to the entire Duetto, which began in C major at 'Sul suo capo aggravi un Dio' (see Example 17). It opens with a

---

[6] Very few 'particelle' exist from the period, unfortunately, especially in Italy. When full sets of material are found, though, from a theater such as the Théâtre-Italien in Paris, 'particelle' are inevitably present.

Moderato theme for Anna, given in Example 31, a transcription of the entire manuscript:

Ex. 31. First version of the cabaletta in the Duetto (No. 7)

The period is simple: two similar four-bar phrases, the first closing in I, the second in III; four bars on V to prepare the cadence; a cadential phrase in I played twice, with a more forceful melodic gesture at the end. Periods of this kind, common in the music of both Donizetti and Rossini, could well be called 'linear' to differentiate them from a typically 'symmetrical' Bellinian period (a a' b a" cadences).[7] The tempo now shifts abruptly to Allegro, and Giovanna's strophe is declaimed to music that resembles a Rossinian 'crescendo', still in C major (mm. 20–36):

$$2 \times (\ 4 \ \overset{\approx}{+} \ 4 \ ).$$
$$I \to V \quad V \to I$$

The second statement of the theme ascends to the high A at its climax (m. 35), but the resolution is interrupted by a deceptive movement to the flattened submediant. For the first time the two protagonists sing together, prolonging the dominant, with Giovanna holding the G and Anna inflecting the music to C minor with the text 'Infelice, non sei rea'. Giovanna alone provides the resolution to the tonic. Donizetti specifies 'Primo Tempo' and Anna repeats her entire period, with two weak interjections added for Giovanna ('No' in mm. 56 and 58). Again the tempo speeds up to Allegro for the concluding cadences notated essentially for Anna alone: $2 \times 8 + 2 \times 2 + (2 \times 1 + 2)$.

The structure of this cabaletta is peculiar in the way Giovanna's strophe is handled. Donizetti supplies Anna with a slow theme, linear and somewhat chromatic. Her passion and abandon have filled the opening section of the Duetto, where she invoked horrendous curses on the head of her unknown rival. Now, knowing this woman to be her closest friend, she masters her emotion and pardons the unhappy Giovanna. Giovanna, on the other hand, reacts to her pardon by becoming hysterical. Her line ascends and descends triadically, while accented dissonance concludes each phrase. Only Anna's phrase is repeated, giving Giovanna's music the structural function of a 'middle section', which its inner shape also seemed to suggest. We are not surprised, therefore, when final Allegro cadences follow immediately and lead in standard fashion to the close. Had Donizetti wanted to underline this interpretation of the structure, he would

[7] Friedrich Lippmann, in his *Vincenzo Bellini*, op. cit., discusses Bellini's typical melodic procedures.

have concluded the piece with an orchestral statement of Giovanna's phrase. This is a structural device employed extensively by Rossini, whereby a musical idea used between the two statements of the cabaletta theme recurs almost inevitably in the orchestral coda to the composition. There is no evidence as to Donizetti's intentions here, however, for the manuscript concludes in the last measure of Anna's part.

Why did Donizetti reject this version? The problem lies almost entirely in Anna's period. Though clearly differentiated from Giovanna's, it is musically quite uninteresting. The C major tonality fights against Donizetti's attempt to color the melody chromatically; the turn to E minor is abrupt and mechanical; and the final cadence, which opens with a staccato arpeggiation of the tonic triad, has almost nothing to do with the sentiment it pretends to express. With a period so generic in quality, it is not surprising that Donizetti sought a new musical setting. In the process, he threw over practically the entire original version, except for part of the cadences, and began again from scratch.

The new version of the cabaletta theme opens with a setting (mm. 1–17) of the first six verses of Anna's strophe. A complete transcription of the second Morgan Library manuscript is given in Example 32:

Ex. 32. Second version of the cabaletta in the Duetto (No. 7)

Anna

gra - zia or chie-do a Di - o, e con - .ces - sa a me sa -

Anna

- rà, sì, con - ces-sa a me sa - rà, la tua gra-zia imploro a

* * : Added measures.

Giovanna

Ah! peg - gio - re è il tuo per -

Anna

Di-o, e con-ces-sa a me sa - rà.

Gio.

- do - no del - lo sde - gno ch'io te - me - a. Pu - ni -
[sic]

Gio.

- tor mi la - sci un tro - no del de - lit - to ond' io son

Gio.

re - a. Là m'at - ten - de un giu - sto Di - o che per

Anna

Ti ri - man - ga in que - sto ad -

Anna: Va, in-fe - li - ce, e te-co

meno

Anna: re - ca il per - do - no di Bo - le - na...

Gio.: Ah! peg-

Gio.: - gio - re è il tuo per - do - no del - lo sde - gno ch'io te -

Gio.: - me - a.

Anna: Ti ri - man - ga in que - sto am - ples - so l'a - mor

Gio.: Là m'at - ten - de un giu - sto

Anna: mio, la mia pie - tà, _____

l'a - mor

Gio.

Anna   Di - o   che per - don   per   me   non   ha, _____

mio, la   mia   pie - tà,   l'a - mor   mio, la   mia   pie -

Gio.

Anna   - tà, _____   ti   ri - man - ga   l'a - mor   mi - o,   l'a - mor

*   * : Added measures.

Gio.   *stringendo a poco, e rinforzando*

Anna   mi - o, la mia pie - tà,   sì,   ti   ri - man - ga   l'a - mor

Gio.

Anna   mi - o,   l'a - mor   mi - o,   la _____   mia   pie -

It continues with the first four verses of Giovanna's strophe (mm. 17–25), after which the two voices come together in a conclusion utilizing the last two verses of Anna's strophe and verses 5 and 6 of Giovanna's (mm. 25–39).[8] The theme is a single entity, then, though within it Anna and Giovanna are at first given distinct melodic phrases. The first eight measures of the new theme are rhythmically and melodically extremely close to those of Example 31. The new tonality, however, makes all the difference, as do some subtle changes between the antecedent and consequent phrases. The declamation of 'La tua grazia or chiedo a Dio' in short, breathless fragments prepares better both the ascent to the high G and the modulation to E minor.

Giovanna's phrase (mm. 17–25) is organically central to the cabaletta theme, transforming the E minor harmony into the dominant of A minor, presumably in preparation for a return to the tonic at the close. Yet although it maintains the insistent motion up and down that characterized the original draft, it has little emotional profile. It merely sits on the dominant, preparing the expected cadences rather than giving Giovanna a dramatic and musical force of her own. Indeed, the neutrality of Giovanna's music is apparent in the reprise of the cabaletta theme, as we shall see, where the organization of the material and its division between the two characters is altered.

In Donizetti's musical universe, the theme of a cabaletta concluding a Duetto whose opening section was in C major cannot be allowed to cadence in A minor. The E major dominant is therefore undercut by bass motion which supports the melodic B with an F natural instead of the anticipated E (m. 25). The resulting dominant seventh chord immediately prepares the cadences *a 2*. The voices open with an imitative passage, then merge together for the final cadences in thirds. Donizetti has thus created a well-constructed cabaletta theme for Anna and Giovanna, with a much more striking opening period and an unusual juxtaposition of A minor and C major. Giovanna's part is not particularly effective, however, nor do the cadences rise above the generic.

---

[8] The final two verses of Giovanna's strophe do not figure explicitly in this version, though it is conceivable, if not likely, that Donizetti intended them to be used in the final cadences, where Giovanna's text is not entered.

The short middle section shows no vocal participation; rather, the bass line ascends alone from C through D and D♯ to E. This last (E) again functions as the dominant, preparing the repeat of the cabaletta theme. This repeat is derived entirely from the original statement, though it is considerably shortened. The structure of this repeat and the way its parts are derived from the original cabaletta theme can be summarized as follows:

Table X. The cabaletta theme and its reprise in the second version of the Duetto (No. 7)

Reprise of the cabaletta theme			Original cabaletta theme	
45–9	Anna	=	1–5	Anna
49–53	Giovanna	≈	5–9	Anna
53–5	Anna	=	17–19	Giovanna
55–7	Anna	≈	21–3	Giovanna
57–71	a 2	=	25–39	a 2

Despite the cuts and the slight alterations in the melodic lines, the passage functions adequately as a reprise of the cabaletta theme. Whatever pretense Donizetti may have had of differentiating the characters in the original statement of the theme, however, disappears completely in the reprise, where the melodic material is assigned to different characters. The final cadences are quite similar to the first version of the cabaletta, but differ particularly in hinting at A minor at the beginning (mm. 71–5). They turn unequivocally to C major only as the voices begin their ascent. Perhaps because of this harmonic ambiguity at the start, Donizetti expands the final measure of the phrase (m. 74 of Example 31) to two full measures (mm. 78–9 of Example 32). He likewise doubles the values in the final cadences (compare mm. 89–90 of Example 31 and mm. 86–9 of Example 32). By altering the declamation of the cadences, finally, Donizetti produces a broader, less hectic cadence.

Though no unequivocal evidence establishes the priority between these two versions of the cabaletta, it seems unthinkable that Donizetti could have discovered the A minor version of Anna's theme and substituted for it the C major version. For better or worse, on the other hand, Giovanna's melody has been completely altered, and the two versions have such scant

Table XI. The three cabalettas for the Duetto (No. 7) of Anna and Giovanna

	1A	1B	1C	1A	1D	1E	2E
First version	1A Anna period Moderato C major	1B Giovanna period Allegro C major	1C Preparation for repeat *a 2* V of C major	1A Repeat of Anna's period C major	1D Long cadence *a 2* C major	1E Short cadences *a 2* C major	2E Short cadences *a 2* C major similar to 1E
Second version	2A Anna period A minor similar to 1A in part	2B Giovanna continuation on V of A minor	2C cadences *a 2* C major	2D non-vocal middle section → V of A minor	2A→C shortened reprise of cabaletta theme A minor→ C major	2D Long cadence *a 2* C major similar to 1D	2E short cadences *a 2* C major similar to 1E
Final version	2A Anna period expanded with new cadence A minor→ C major Moderato	1B Giovanna period C major Più mosso	1C Preparation *a 2* on V of C major	New passage *a 2* voices in alternation at two-measure distance C major	1D Long cadence *a 2* C major		2E Short cadences *a 2* C major

(Second version: bracket marked "cabaletta theme" spanning 2A and 2B.)

relationship to one another that arguments about the different qualities of the settings are irrelevant: the two are separate inspirations.[9] Both versions of the piece resemble traditional cabaletta designs, though neither is absolutely standard. Each has positive characteristics the other lacks.

For his definitive version, Donizetti was to adopt the strongest elements in each setting. Table XI shows schematically how Donizetti derives the final version of the cabaletta from his two earlier efforts (see p. 123).

With only minimal changes, the composer takes Anna's A minor theme from the second version. He first includes, then ultimately crosses out the two additional measures appearing in the margin of the Morgan manuscript (Example 32, mm. 13–14). For the last two lines of Anna's strophe, Donizetti provides a new cadential passage (see Example 33(a)), particularly effective in its fluctuation between A minor and C major. Since the cadences point back towards A minor, the turn to C major in the fourth measure of Example 33(a) seems at first a mere detail, but at the close of the period Donizetti shifts the tonal center definitively to C major, arriving on the new tonic as Giovanna's period begins. This section in C major for Giovanna is derived without change from the first version. Having allowed Anna's period to close with a cadence in C major, Donizetti could remove Giovanna from the composite theme of the second version and allow her to preserve the well-articulated period he had originally given to her. He follows the first version through the preparation for an anticipated reprise, but eliminates the final four measures of this section (Example 31, mm. 44–7), which had resolved the dominant harmony and introduced the repeated cabaletta theme.

Poised on the dominant of C, committed to two distinct periods for his characters which begin, respectively, in A minor and C major, Donizetti decided to avoid the problematic reprise altogether. Instead, he remembered the close of the cabaletta theme in the second version (Example 32, mm. 25–39), where the voices began to weave imitative patterns before joining together *a 2* for the cadence. Though he changes some melodic detail, he

---

[9] I have treated a rather similar chronological problem in 'The Arias of Marzelline', op. cit., though in that case manuscript evidence was able to buttress the musical argument.

Ex. 33. Two added passages in the definitive version of the cabaletta in the Duetto (No. 7)

(a) Cadential phrase for Anna's strophe

(b) Added passage, with the voices in imitation

essentially adopts these measures, prefacing them with another passage alternating the voices in two-measure segments (see Example 33(b)). Thus, a lovely progression is achieved, gradually bringing the voices in closer imitation until they sing together. The concluding cadences, which draw in part on both earlier models, remain *a 2*. Donizetti chooses the first version (Example 31, mm. 67–74) for the initial cadential phrase. Presented in the Morgan manuscript within repeat signs, he copied it the same way in his autograph, only later crossing these out. The second version (Example 32, mm. 80–90) provides the remainder of the cadences. In this final version, where C major reigns alone from Giovanna's period through the close, the recurrence of an A minor inflection, as in the initial cadence of the second version, would have been misplaced. That is why in the eight-measure cadential phrase Donizetti returns to his first conception.

Irrespective of its origins, the cabaletta of the Duetto for Anna and Giovanna is the most formally innovative cabaletta movement in *Anna Bolena*: flexible in its contents, inventive in its use of tonality, striking in the way the voices gradually come together for the final cadences. It is a model to which later composers, including Donizetti himself, could effectively turn. A Verdi, for example, would find in this cabaletta the 'dissimilar' handling of 'similar' poetic strophes. Yet the piece is not without problems, and the primary problem betrays its composite origins: the sections lack an organic internal relationship. Despite detailed changes and the addition of extra music preceding it, the passage that had functioned as a cadence to the cabaletta theme in the second version (2C in Table XI) does not have enough melodic interest to sustain the latter half of the cabaletta. After the middle section on the dominant, there is nothing but a succession of dominant and tonic harmonies for measures on end. The insistence on C major in this section functioned well when it was the cadence of the cabaletta theme in the second version, for there it had to counter-balance an opening entirely cast in A minor. In this final version, however, the insistence on C major has little musical justification, especially since Giovanna's strophe, taken from the first version, is entirely constructed of tonic and dominant harmonies in C major; indeed, the ubiquity of these two harmonies in the final version leads the music into banality.

The problem is not one of daring but rather of realization. Donizetti saw the potentialities of the material and recognized the weaknesses of both early versions. He unquestionably took the best parts of each, and the cadence he wrote for Anna's strophe is excellent. But having abandoned the repeat of the cabaletta theme, he was at a loss to find a replacement. Abstractly, the gradual merging of the voices is a fine idea, one that serves the drama well, but the actual musical setting does not sustain our interest. One must regretfully consider this cabaletta a failure, but in its failure it is far nobler than many victories. Just how much nobler it is, we can learn by examining one additional cabaletta in *Anna Bolena*, that of the Terzetto.

(d) Terzetto

The Terzetto, 'Ambo morrete, o perfidi', is certainly one of the great moments in *Anna Bolena*, though it is unmistakably Rossinian in design. We have seen that its cabaletta excited both rapturous praise and vehement denunciation. Indeed, it is hard to imagine that the various reviewers are discussing the same piece. They may not be, though from the viewpoint of these critics the textual problems about to be discussed would probably not have altered their stance. The cabaletta for the Terzetto exists in two separate forms, a 'cabaletta vecchia' and a 'cabaletta nuova'. Presumably the change was one of those made during the course of the first season.[10] Indeed, some or even all the changes in the

---

[10] See the allusion to these changes in the Milanese paper, *L'eco* of 4 February 1831, cited above (p. 5).

autograph involving completely orchestrated music may very well derive from these mid-season adjustments.

Table XII summarizes the state of the autograph manuscript for the conclusion of the Terzetto:

Table XII. Cabalettas for the Terzetto (No. 8(b)) in the autograph

Layer	Folios	Comments
1	304–305v	Enrico's opening theme (identical in the two versions);
		Ten measures of the Anna—Percy continuation in the old version. These measures are crossed out, and over them Donizetti has written 'alla stretta nuova'. But he (?) has also written 'si fa tutto'.
3	306–315v	A copyist's manuscript of the remainder of the old cabaletta, continuing through the measure preceding the final cadences, 'Più mosso', identical in the two versions.
2	316–320v	Autograph of the 'stretta nuova', picking up after Enrico's opening theme, complete through the measure preceding the final cadences.
1	321–324	Autograph of the final twelve measures of the old cabaletta, identical to those found at the end of the copyist's manuscript;
		Autograph of the cadences common to the two versions.

Though the autograph of much of the old cabaletta is missing, we can reconstruct both versions. Among the various states of the first, oblong Ricordi edition, furthermore, edited versions of both cabalettas exist, although in the later, upright edition, only the new cabaletta appears. Indeed, the states of the 'first edition' are interesting enough to warrant consideration here.

These early, oblong editions of the operas of Rossini, Bellini, Donizetti, and Verdi, not to mention other composers who worked with the house of Ricordi, demand intensive scholarly investigation. Each impression, or press run, produced only a small number of copies, so that surviving copies of a given opera tend to exist in many different states, often with individual pages re-engraved, as plates wore out, developed cracks, etc.[11] The history of an edition can often be traced by observing the initials adjacent to the plate number, initials originally used to identify the specific engraver responsible for the engraving of a particular page. Usually these matters are of bibliographical interest alone, not significantly affecting the musical text, but in the case of 'Ambo morrete, o perfidi' and its cabaletta, the stages of the first Ricordi edition help us understand the history of the work.

The Biblioteca di Santa Cecilia in Rome possesses three copies of this edition (Copy A: 10.A.28; Copy B: 5.B.17; Copy C: 102.B.17). By comparing the engravers' initials it is possible quickly to order the editions chronologically. If we take the first nine pages of the edition, for example, we arrive at the following chart:

page	Copy A	Copy B	Copy C
1	[E]C	ZZ	ZZ
2	EC	EC	AA
3	EC	ZZ	ZZ
4	FF	AA	AA
5	FF	FF	AA
6	FF	FF	aa
7	O e B	ZZ	ZZ
8	O e B	AA	AA
9	O e B	AA	AA

It is almost invariably the case that plates with initials standing for real names (such as EC or O e B) precede plates with more generic initials (such as ZZ or AA). In the earliest of these three copies, unquestionably a very early state of the edition, EC engraved pages 1–3, FF pages 4–6, and O e B pages 7–9. As we might expect, there is a logic to the order of engravers' initials in early states of an edition that disappears as plates are successively replaced. When Copy B was run off the press, only pages 2, 5, and 6 were printed from the same plates; all three of these plates had in turn been replaced by the time Copy C was issued. Similar patterns are found throughout the score, and nowhere is the postulated order contradicted by the engravers' initials.

[11] A general discussion of these problems in printed musical editions is to be found in D. W. Krummel, compiler, *Guide for Dating Early Published Music: A Manual of Bibliographical Practices* (Hackensack, New Jersey, 1974), pp. 30 ff.

Though in pages 1 to 9 no significant changes were introduced during re-engraving, the same cannot be said for the music of the Terzetto cabaletta preserved in these three states of the first edition. Copy A, which is closest to the original state of the edition, includes not the 'old cabaletta' of the Terzetto, as we might expect, but the 'new' one, although it does not so identify it. Copy A, however, makes an absolutely absurd error in the cabaletta. While composers may not normally have proof-read these piano-vocal scores with any real care, if at all, this error is so egregious that surely Donizetti or someone else must have brought it to Ricordi's attention. The repetition of the cabaletta theme (as we shall see, the 'new cabaletta' is formally more conventional than the 'old') is a shortened version of the theme itself. Enrico sings the first half of his opening period, up to the arrival on E major at 'Il tuo nome, il tuo sangue sarà'. Donizetti then jumps to the second half of the contrasting theme sung by Anna and Percy. Through the end of this shortened repetition, Donizetti provides an abbreviated score in the autograph. He writes explicitly only the bass line and an occasional violin figure; the remainder of the orchestral parts and the vocal lines for Anna and Percy are to be read from the original statement of this material. To the original texture, however, Donizetti also adds an explicit new part for Enrico. Thus, the repetition of the cabaletta theme concludes impressively *a 3*, where it was *a 2* before. Though there is no question whatever that Anna and Percy's parts must be included in the general repetition of the earlier passage, in the first state of this Ricordi first edition their parts were omitted, leaving Enrico to sing alone during the entire abbreviated repetition of the cabaletta theme.

At roughly the same time that Ricordi became aware of the error, Donizetti must have come to feel that both versions of the cabaletta were worth preserving in print (as the state of the autograph also attests), but the nature of the changes made in the edition are rather curious. In the first state, the plate number '5138' was used for the entire Terzetto, including the new cabaletta; when the score was re-issued, it became available in either of two versions, one with the 'cabaletta vecchia' of the Terzetto, the other with the 'cabaletta nuova'. The plate number '5138' is now used for the 'cabaletta vecchia', and the plate number '5138/1' is adopted for the 'cabaletta nuova'. Copies B and C at the Biblioteca di Santa Cecilia preserve, respectively, these states of the edition. The 'new' cabaletta is the same as in Copy A and in modern scores of *Anna Bolena*, but with the error of Copy A corrected, Anna and Percy joining Enrico at the end of the reprise; the 'old' cabaletta is the piece we know from Donizetti's autograph score.[12]

Both cabalettas begin with the same extended period for Enrico, 'Salirà d'Inghilterra sul trono', and both conclude with the same cadences *a 3*. Between these poles, the settings are very different. The 'new' cabaletta continues with a lengthy period for Anna and Percy, sung almost entirely in unison. Like Enrico's period, it is in C major and it has an internal modulation to the third degree. A formal 'middle section' is present, shaped as a Rossinian crescendo in the orchestra, with superimposed, overlapping vocal lines:

$$3 \times (\quad 4 \quad \overset{\approx}{\mp} \quad 4 \quad).$$
$$\text{I} \rightarrow \text{V} \quad \text{V} \rightarrow \text{I}$$

The music continues to sit on the tonic, preparing a shortened reprise of the cabaletta theme. Employing the cadence on the third degree common to the two melodic periods, Donizetti presents the first half of Enrico's period and the second half of the period for Anna and Percy, adding a new part for Enrico at the close, as mentioned above. A series of straightforward cadences brings the cabaletta to a close.

The structure of the 'old' cabaletta is much more unusual than its revision. Example 34 gives a reduction for piano and voices of the music which the two versions do not share, that is, everything between Enrico's opening period and the final cadences. Like the 'new' cabaletta, the 'old' one contrasts the themes sung by Enrico and by Anna and Percy, but unlike it, this version avoids all formal restatement of the cabaletta theme, even if Enrico's part at the end is similar to the close of his initial period, 'Salirà d'Inghilterra sul trono'. There are some wonderful moments in this first version of the cabaletta: the sudden

---

[12] That the 'old' cabaletta is indeed the first version is further attested to by the original libretto, which gives as Anna's first line in the cabaletta: 'Quanto, ahi quanto! è funesto il tuo dono.' The expletive 'ahi' is present in the 'old' cabaletta, but missing in the 'new' one.

Ex. 34. The 'old cabaletta' of the Terzetto (No. 8(b))

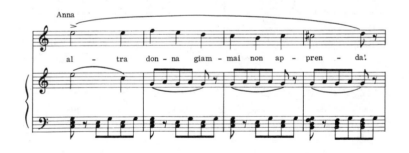

Per. al - tra don - na giam - mai non ap - pren - da.

Enr. u - na don - na più de - gno d'af - fet - to,

Per. L'In - ghil - ter - ra giam - mai non ap - pren - da

Enr. e il tuo no - me in - fa - ma -

Per. l'em - pio stra - zio che d'An - na si fa,

Enr. - to da tut - ti sa - rà, sa -

modulation to E♭ major, the marvellous descending and ascending chromatic passages with the high G pedal ringing out for the
indefatigable Rubini, the immediate juxtaposition of the duet for
Percy and Enrico and the previous held seventh chord on G *a 3*,
and the superb cadential phrase (which with small changes is also
used in the 'new' cabaletta). But there are some real infelicities
too: the phrase shared by Anna and Percy is terribly flaccid, and
must be rescued by Enrico's dramatic entrance on 'Salirà'; the
wonderful chromatic ascent to the held dominant seventh chord
is vitiated by twenty additional measures, based on that same
Anna and Percy phrase, which continue for too long the G pedal.
The 'old' cabaletta, however, is certainly interesting enough to
justify Donizetti's decision to preserve it.[13]

The 'new' cabaletta, despite its more conventional external
form, is probably the more effective composition. The unison
tune for Anna and Percy, a texture that can often descend into
banality, is here extraordinarily exciting. But even the 'new'
cabaletta is marred by its middle section, which is melodically and
rhythmically undistinguished, especially in the very weak harmonic preparation for the repetition of the theme (alternation of

[13] It is worth noting that the lengthy cadence opening the cadential section,
3 × 2 + 18, was originally enclosed in repeat signs. It is not clear when Donizetti
crossed them out, though it may have been when he decided to change the
cabaletta.

the tonic harmony with a diminished seventh chord also built on C), which scarcely provides any impetus for the repetition.[14] Neither the 'old' nor the 'new' cabaletta, then, is ideal, but each has some excellent qualities. It is particularly poignant that Donizetti's attempt in the 'old' cabaletta to avoid the Rossinian formal model comes so close to succeeding, but ultimately cannot sustain itself. The 'new' cabaletta, for all its obvious strength, is practically a cry of defeat.

All these examples represent diverse attempts by Donizetti to adapt the conventional cabaletta design to new dramatic and musical purposes. At this moment in his life, the composer does not presume to question the presence of a cabaletta at the close of each musical number: it is axiomatic to his musical style. But he is constantly at war with the traditional cabaletta design, prodding here, pushing there, cutting, revising. The same tensions we have observed before appear here in a particularly extreme form. The Rossinian model is so clear, so perfect in its way; yet its repetitiveness, its invitation to ornament the melodic line, its way of absorbing different characters of an ensemble within a single musical framework, all these factors symbolized a conception of opera against which composers such as Donizetti and Bellini were reacting. What is fascinating about the final versions of these four examples is that they provide symbolic models for the ways composers could come to grips with the problem of the cabaletta: adapting its structure to specific dramatic forces (Finale 1°), cutting the traditional repeat (Quintetto), searching for a new formal design (Duetto Anna e Giovanna), or modifying the traditional design (Terzetto). In every case, the final version was the product of extensive revisions, revisions that underline as clearly as possible the ambiguous attitude of Donizetti to the cabaletta. This unhappy mixture of attraction and scorn would continue to plague Italian composers for the next fifty years.

[14] Then again, even Verdi faced with the repetition of the cabaletta theme rarely knew how to make the musical gesture dramatically meaningful. For every 'Sempre libera' there are twenty 'O mio rimorso! oh infamia!' (the cabaletta from Alfredo's 'De' miei bollenti spiriti' in *La Traviata*).

X. *The Interdependence of Compositional Decisions*

PROBLEMS of detail and problems affecting the cabaletta merge in a striking fashion in the first-act Duetto for Giovanna and Enrico (No. 2). Although we do not basically deny Barblan's critical judgement that this piece 'does not surpass the usual formulas of good melodramatic theater of the early nineteenth century',[1] in at least one respect the Duetto is unusual: the explicit melodic relationship Donizetti establishes between the *cantabile* second section and the concluding cabaletta. Melodic relations between the first section and the *tempo di mezzo* are typical of Rossini's duets, but melodic ties between lyrical sections are practically unknown, and the device helps give this Duetto a character of its own. In Donizetti's original conception, however, the relationship is not found, and the history of its inclusion is worth following.

The opening section of the Duetto begins with a standard, formalized confrontation.[2] Often in a Rossini duet each character sings an introductory passage and a main period; the introductory passages are different, while the main periods are practically identical. Here Donizetti not only differentiates the introductory passages, but further recasts much of Giovanna's vocal line in the main period, leaving unchanged the orchestral accompaniment. Such melodic 'variation' has often been invoked in discussions of Donizetti's style, as mentioned above, and here the composer uses it well to help differentiate his characters.

Each solo statement, Enrico's and then Giovanna's, originally contained eight additional measures of cadential material (essentially 4 ╪ 4), as in Example 35 (see pp. 142–3). The absence of these cadences in the finished score is very noticeable, since a textual disturbance is thereby created that Donizetti *as composer* would

---

[1] Barblan, *L'opera di Donizetti*, op. cit., p. 52.

[2] See my article, 'Verdi, Ghislanzoni, and Aida', op. cit., which is largely concerned with the standard Rossinian duet.

Ex. 35. Deleted cadential passages from the first section of the Duetto, Giovanna and Enrico (No. 2)

(a)

(b)

not have permitted himself, but that *as reviser* he could accept. Enrico's original stanza reads as follows:

> Fama? Sì: l'avrete, e tale
> Che nel mondo egual non fia:
> Tutta in voi la luce mia,
> Solo in voi si spanderà.
> Non avrà Seymour rivale,
> Come il sol rival non ha.

After using the first two verses in his introductory passage, Enrico adopts the next four for the main period. He sings them once, then repeats them. By cutting the eight-measure cadential phrase, though, Donizetti also omits the penultimate verse, allowing the final verse to stand without its antecedent, 'Non avrà Seymour rivale', which alone gives meaning to 'Come il sol rival non ha'. Since this is a repetition of the text, we understand the basic sense, to be sure, but the cut introduces a textual abnormality. Similar remarks could be made about Giovanna's cadential phrase.

Both these passages are fully orchestrated; hence, Donizetti's decision to omit them came late in the history of the piece. Why did he choose to let them go? One could answer that he did it for the sake of brevity, but rarely does Donizetti wait until music is orchestrated before making that kind of judgement.[3] The situation is further complicated by the fact that in the *cantabile* section, too, Donizetti crosses out a completely orchestrated cadential repetition, affecting precisely the cadential phrase that links the *cantabile* and cabaletta of the Duetto in its final version. Giovanna first sings this phrase alone, to the text 'Di un ripudio avrò la pena, né un marito offeso avrò'. After a few measures, she sings it again with Enrico. This statement of the phrase *a 2* was

Ex. 36. Deleted cadence from the second section of the Duetto (No. 2)

⊛→ ←⊛ : Deleted bars.

[3] There is an example of just such a cut 'for brevity' here. Before the main part of Giovanna's period ('e quell'ara è a me vietata'), Donizetti originally planned an additional eight measures. Though he writes only the bass part, it is clear that this was an introductory orchestral statement of the main theme, similar to the one that occurs at the same point in Enrico's period. The measures were subsequently struck out. We cannot be sure when he made this cut, since all the orchestration of this part of Giovanna's period is read 'come sopra' from earlier material, and hence all of it is given with the bass line alone.

originally repeated yet again by Donizetti, with different harmonic details and a slightly fuller orchestration, marked by a lovely chromatic inner voice (see Example 36). In order to understand this omission, as well as the cadential cuts in the first section of the Duetto, we must turn to the cabaletta.

The cabaletta cost Donizetti much trouble, and not until the very last stages does he actually work out its melodic reference to the *cantabile*. Here is his first attempt at a cabaletta theme:

Ex. 37. First version of the cabaletta in the Duetto (No. 2)

It begins on f. 78r–v, which was originally followed directly by ff. 82–3 (ff. 79–81 belong to a later layer). The manuscript of this first attempt contains, for voice and bass alone, a complete cabaletta theme with dissimilar musical periods for Giovanna and Enrico, though they have parallel poetic stanzas:

Giovanna   Ah! qual sia cercar non l'oso
 No'l consente il core oppresso . . .
 Ma sperar mi sia concesso
 Che non fia di crudeltà.
 Non mi costi un regio sposo
 Più rimorsi, per pietà!

Enrico   Rassicura il cor dubbioso;
 Nel tuo Re la mente acqueta . . .
 Ch'ei ti vegga omai più lieta
 Dell'amor che sua ti fa.
 La tua pace, il tuo riposo
 Pieno io voglio e tal sarà.

Giovanna first sings verses one to four of her stanza; Enrico does likewise, though to a totally different melodic line. The characters join for a passage *a 2* that concludes the cabaletta theme, each character singing the final two verses of his stanza.[4] Giovanna's contribution to the cadential phrase, with Enrico in triplets below, is similar in character to the cadential theme of the *cantabile*, though different in realization. Perhaps the similarity ultimately caught Donizetti's attention. In any event, the theme is brought to a conclusion, with some uncertainty at the end,[5] and is followed by two additional measures, subsequently crossed out, clearly beginning the middle section of this cabaletta. The manuscript breaks off here.

This first version did not meet Donizetti's needs—Enrico's eternal triplets, dull and predictable, may be partly responsible—and the composer decided to alter the design. He kept f. 78r–v, but crossed out the final measure, which had resolved Giovanna's

[4] In the final version, the last two verses of Enrico's stanza are not set. Once again, the original libretto informs us about earlier versions of the music.

[5] The most significant change occurs between 26 and 27, where Donizetti planned a two-measure cadence, similar to the three-measure cadence that concludes the theme. The final cadences, in short, were originally $(2 \widetilde{+} 3)$, with the typical Donizettian expansion for the last cadence. The first element was then eliminated.

opening phrase and begun Enrico's part. He then inserted the present f. 79 (a single folio) and ff. 80–1 (a bifolio). These add a lengthy cadential phrase for voice and occasional bass to Giovanna's opening theme (incorporating verses five and six of her stanza). After a rapid modulation to A major in the bass, Enrico repeats the theme in the dominant (though the manuscript ends before he gets through it all). The second version of the cabaletta is given in Example 38, picking up from the last measures of Giovanna's part in Example 37 (mm. 8–9):

Ex. 38. Second version of the cabaletta in the Duetto (No. 2)

This is the most typical cabaletta construction for a duet, with each character singing the same phrase, though the tonal contrast is not obligatory. The use of the dominant, however, practically assures us that a brief middle section will prepare a repetition of the theme *a 2* in the tonic. Donizetti's motivations for rejecting this version seem clear: the theme is terribly long and undistinguished; the cadential sequence is depressingly regular; and all hints of the cadential relationship to the second section of the Duetto are gone. Donizetti may never have written any more of this version: the prospect of the inevitable reappearance of his theme may have deterred him.

In neither of these cases did the composer write more than the melodic lines and barely adequate bass. Now, using the same manuscript paper but crossing out both earlier versions, Donizetti prepared the definitive cabaletta. It is found on ff. 78r–82v. It resembles more his first version than his second, and avoids the latter's symmetry by immediately differentiating the characters. One of the definitive phrases (Giovanna's 'più rimorsi, per pietà') was originally proposed for 'Che non fia, che non fia di crudeltà', with the same harmonic motion to F♯ (see Giovanna's opening phrase in Example 37). Otherwise, the definitive theme is less florid than the earlier versions. Most significantly, Donizetti brings the two voices together in a fine ascending, chromatic sequence that leads into the cadential phrase, for the first time an exact reference to the cadences of the second section (given in Example 36). The effect is superb and caps the Duetto in a way Verdi would emulate many years later in the first-act Duetto for Lady Macbeth and Macbeth.[6] A short middle section now leads to an abbreviated repetition of the cabaletta theme (including, of course, the by now familiar cadential passage) and a concluding cadential section.

Finding the right musical gesture for the cabaletta cost Donizetti great pains, and in the autograph we can watch the evolution of all three versions. The ultimate writing came quickly, with no signs of hesitation, not even at the quotation from the *cantabile* section of the Duetto. Still, it should be abundantly clear that the most characteristic gestures, those that interpret conventional procedures in new ways, are often won only through the most arduous compositional labors.

The cuts observed in the cadences of the first section and the *cantabile* of the Duetto (see Examples 35 and 36) surely result from Donizetti's decision to highlight so prominently the cadential phrase linking the *cantabile* and cabaletta. For all its beauty of realization, after all, the cadence remains a simple I–IV–V–I design. Donizetti may have also felt that the equally simple I–ii–V–I cadences closing each solo in the first section of the Duetto were not only unnecessary, but would actually weaken the later, more important figure. Likewise, his deletion of a third statement of the repeated cadential phrase in the *cantabile* probably resulted from the desire to avoid overexposure of this material, once its reappearance in the cabaletta was determined. The smaller cuts, in short, reflect the larger context. They help make this Duetto not so much a piece that goes beyond Rossinian conventions (nothing formally in *Anna Bolena* can be said to do that), but rather a particularly apt and personalized adaptation of those conventions.

---

[6] Significantly, though, the most striking repetition of the cadential phrase, between the first and final sections of the duet, was not present in Verdi's original, 1847 version. It is a feature of the Parisian revision of 1865. See Budden, *The Operas of Verdi*, vol. i, p. 288.

XI. The 'Duetto Rifatto' for Anna and Percy (No. 5(b)')

IN order to demonstrate the variety of compositional decisions Donizetti made while composing *Anna Bolena*, we have thus far organized our examples by types of musical problems: melodic detail, continuity, the cabaletta, etc. It will prove useful now to change our perspective and to study instead all the evidence associated with Donizetti's work on a single musical number. The newspaper report informing us that Donizetti made several important changes in *Anna Bolena* during the course of its first season does not specify what these changes were. Nor can we turn to printed librettos for further information, since these librettos, normally published at the beginning of a theatrical season and reflecting the version of an opera at that moment, were rarely if ever reprinted to reflect changes introduced during the course of the season. Examination of Donizetti's autograph does allow us to make some likely suppositions. The shortening of the opening chorus in the Introduzione is one candidate; the substitution of the 'cabaletta nuova' for the 'cabaletta vecchia' in the Terzetto is another; the replacement of the cabaletta in Percy's Cavatina is at least possible. The most important change, however, was the addition of a completely new Duetto, 'Sì, son io che a te ritorno' (No. 5(b)'), replacing the original piece, 'S'ei t'abborre, io t'amo ancora' (No. 5(b)), at the start of Finale 1°. The modifications visible in the autograph of this piece are particularly interesting and summarize neatly most of the points we have made so far.

In the dramatic design of the opera, this sole direct confrontation between Anna and Percy is crucial. Percy, the love of Anna's youth, has been recalled from exile to the English court by Enrico, who hopes thereby to bait a trap in which his Queen will forget herself, allowing Enrico to dissolve their marriage and wed Giovanna Seymour. The original Duetto and its extremely dramatic introductory recitative seem to have pleased neither the public nor the singers, for during Donizetti's life they were frequently reduced to a short recitative or replaced altogether by the new Duetto, which is bound into the autograph at the close of Act I. Its text, which incorporates the action of both the recitative and the original Duetto, is based on the premise that Percy enters believing it was Anna herself who persuaded the King of his innocence and therefore was responsible for his being recalled from exile. But his youthful enthusiasm quickly changes as he witnesses the depth of Anna's sorrow. He becomes bolder in his desire to assist her. Though Anna continues to refuse his love with her words, she finally abandons herself in his arms. With the reappearance of Smeton and entrance of Enrico and the court, the trap is sprung.

Let us begin with the text of this Duetto. Although we have no proof, there is every reason to think the text was arranged for Donizetti by his original librettist, Felice Romani.

('Chiamata: *Anna*:   Riccardo!')

Moderato

Riccardo	Sì, son io che a te ritorno
	Nel pensier di lieta sorte.
	Te perduta, al cielo un giorno
	Io chiedea, chiedea la morte.
	Or che a te mi vuoi vicino
	Chiedo al ciel miglior destino;
	Ai ridenti giorni anelo
	Della nostra prima età.
Anna	Sciagurato, ignori forse
	Che sei tu d'Enrico in Corte.
	Dell'amor l'età trascorse,
	Or qui regno ha infamia e morte.
	Non io chiesi il tuo ritorno.
Riccardo	No?
Anna	Fu il Rè . . .
Riccardo	D'esso . . .
Anna	e per mio scorno!
	Vanne e salva il mio decoro;
	Te pur salva per pietà.
Riccardo	Io lasciarti? . . . e tu mel dici? . . .
	Ah! crudele!
Anna	Fuggi! Va!
	(*Percy la prende per mano*)

Larghetto

Riccardo Per vederti invidiata,
    Sol per darti onore e fama,
    Questo misero che t'ama
    Altrui cederti poté.
    Ma in trovarti sventurata,
    Il mio dono ancor riprendo,
    E da te, da te pretendo
    L'amor primo, la tua fé.

Anna Oh Percy, nemico è il fato;
    Qui si tesse orribil trama;
    Qui una vittima si brama:
    Spetta forse addurla a te.
    Non fia duol morirti a lato,
    Ch'io con te morrei contenta;
    Ma l'infamia mi spaventa,
    Questa sol fia grave a me.

Allegro

Riccardo Disperati i giorni tuoi
    S'ei t'abborre qui vivrai.

Anna M'ami tu?

Riccardo   Sì.

Anna   Ebben, non puoi
    Me infelice far giammai.

Riccardo T'odia Enrico.

Anna   Io moglie sono . . .

Riccardo D'un perverso!

Anna   Del tuo Ré!

Moderato—Più mosso

Riccardo (*deciso*) Ebben!
    Restati, pur m'udrai
    Spento ma a te fedel.
    E allor rammenterai
    Che fosti a me crudel.

Anna Vivi, mio ben, m'udrai
    Spenta ma a te fedel.
    E allor rammenterai
    Che fui con me crudel.
    (*Qui si abbandona fra le braccia di Percy*)

As one can see at a glance, the organization of the text follows the standard Rossinian model for a duet. The first section consists of parallel stanzas for Percy and Anna (apart from the brief interruptions of Percy during Anna's stanza).[1] Some additional verses lead to the *cantabile*, where we again find a set of parallel strophes for the two characters. A brief dialogue introduces the cabaletta. Only in these last parallel strophes do the verses, *ottonari* up to this point, change to *settenari*. This poetic model, in which the metrical structure of the verses changes only as the cabaletta begins, was widely adopted in Italian librettos of the nineteenth century. As late as 1870, writing the libretto of *Aida* for Verdi, Ghislanzoni turned to this structure for the Duetto of Amneris and Radames, where verses in *settenari* become *ottonari* as the cabaletta begins.

Parallel strophes, parallel music: that was certainly the archetypical musical practice of the period. Yet we have already studied examples in both Rossini and Donizetti where the specific realization of the archetype favored a more vital dramatic interaction. Two techniques were of particular importance in the first section of a duet. Faced with parallel strophes of eight verses apiece, both composers, instead of assigning the same music to the characters for the entire strophe, frequently divided the text into two parts. The first four verses of each stanza were set independently; only for the final four verses and cadences would the periods be essentially identical. The first section of Rossini's Duetto from *Semiramide*, 'Ebben, a te, ferisci', is constructed in this manner, as is the original Duetto for Anna and Percy in *Anna Bolena*, 'S'ei t'abborre' (No. 5(b)). Equally important for Donizetti, though not for Rossini, was the technique of written-out melodic variation, where throughout essentially identical passages the melodic line is significantly modified for reasons of tessitura and also for dramatic effect. As we have seen, Donizetti mixes both these approaches in setting the parallel strophes that initiate the Duetto for Giovanna and Enrico (No. 2).

In preparing the new Duetto for Anna and Percy, Donizetti employed the technique of melodic variation for the entire first

---

[1] At least one of these interpolations, probably 'D'esso', was supplied by Donizetti, for the line does not scan properly in metric terms: it has one syllable too many. It seems very unlikely that Romani or another competent librettist would have permitted himself this irregularity.

section. Percy enters with a melodic phrase full of passion and life, with lovely, fresh ornaments and without important dissonance (except for that which colors the word 'morte'). Here is Percy's entire period:

Ex. 39. Percy's opening period in the new Duetto (No. 5(b)')

-sti - no; ai ri - den - ti gior - ni a - ne - lo del - la

no - stra pri - ma e - tà, ai ri - den - ti gior - ni a -

- ne - lo del - la no - stra pri - ma e - tà, sì, del - la

no - stra pri - ma e - tà, sì, del - la

Donizetti had no difficulty whatsoever in the first part of this period, if we can trust the evidence of the manuscript. Only in two cadential phrases do we find slight alterations, both of which add graceful, ingenuous florid passages that help the composer express in musical terms the almost naïve character of Percy. In place of mm. 20–3 of the final version, Donizetti originally wrote only two bars, which he later enclosed in repeat signs (Example 40(a)):

Ex. 40. Two altered passages in Percy's opening period in the new Duetto

(a)

* The repeat signs are a later addition.

(b)

He subsequently crossed out the vocal line, providing not only
two new measures for the voice but also a variant for the
repetition, and rewrote the orchestral parts as well. The vocal
leaps and accents, the expanded register, the internal variation, all
these factors provide the revised version with a sense of life and
freshness the original lacked. The composer also altered the final
cadence, the original version of which is found in Example 40(b).
In his definitive version, Donizetti replaced the melodic iteration
with a more impassioned design. These are small changes, to be
sure, but in helping to characterize Percy they are efficacious.

As he confronted Anna's period, Donizetti began instinctively
to prepare a melody practically equivalent to that of Percy
(Example 41(a)):

Ex. 41. Melodic variation in Anna's parallel period in the new Duetto

(a)

(b)

(c)

(d)

Almost immediately his sense of the words and his awareness of the need for melodic variation induced him to alter it as in Example 41(b). From that moment on, the principle of dramatic variation was fixed for the entire period. Where Percy sang the phrase given in Example 41(c), Anna responded with that of Example 41(d). The rhythm of the initial figuration is changed and, by shifting the register at 'infamia e morte' while introducing chromaticism, Donizetti gives a different stamp to the entire melody and hence to the characters. The florid writing, so illustrative of Percy's sensibility, has almost disappeared entirely, replaced by a much more syllabic and dramatic declamation. As for the final cadential phrase of which we spoke earlier (mm. 20–3 of Example 39), Anna at first was assigned precisely the same melody as Percy (Example 40(a)), to the words 'te pur salva per pietà'. Taking the opposite approach to the one he adopted for the part of Riccardo, which he made more florid, Donizetti gave that of Anna greater force (Example 42(a)):

Ex. 42. Melodic variants in the cadence of Anna's period in the new Duetto

(a)

(b)

At the final cadenza, to which Donizetti had added a lovely melody for Percy, he left Anna a written declamation that seems almost brutal (Example 42(b)). All the examples of changes in detail and melodic variation present in *Anna Bolena* and other scores of Donizetti very nearly find in this first section of the new Duetto for Anna and Percy their quintessential exemplification. The composer transforms the rather formal encounter of the archetypical Rossinian duet into a confrontation of characters burning with passion.

The two verses that bring this section to a close,

Riccardo Io lasciarti? . . . e tu mel dici? . . .
     Ah! crudele!

Anna   Fuggi! Va!

may not have been present in the text originally submitted to Donizetti. Consequently the music proceeded directly from the first section to the *cantabile*, as in Example 43(a).

This version was dramatically weak, for it left no interval between the lyrical periods. Musically similar cadences and orchestral introductions are precisely the type of passages Donizetti omitted wherever he could, as he preferred a musical continuity less formal, more intense. We should not be surprised

Ex. 43. The transition to the *cantabile* in the new Duetto

(a)

(b)

to learn, therefore, that he crossed out these measures, found at the end of f. 211v and at the start of f. 213r, and inserted instead a new folio (f. 212) containing a short interchange between the characters (Example 43(b)). Though these transitional measures do not completely fill what is unquestionably a gap in the dramatic and musical shape of the Duetto, they are certainly superior to the original version. Only now does Donizetti pass, without an orchestral introduction, to the Larghetto *cantabile*.

In the Larghetto both characters are given the same period. The few melodic variants between their strophes reflect small differences in the accentuation of their respective texts. Though Donizetti wrote the melody without significant corrections, one interesting variant affects the cadence of Percy's phrase, originally equal to that of Anna's (Example 44(a)):

Ex. 44. A cadential revision in the *cantabile* of the new Duetto

(a)

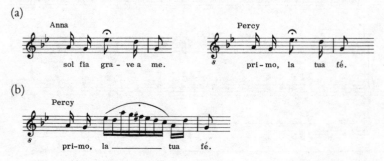

(b)

In his definitive version, Donizetti added a florid cadenza (Example 44(b)), again employing coloratura to develop better the personality of Anna's youthful lover. The melodic period of this Larghetto is one of the most beautiful in all Donizetti's music:

Ex. 45. Percy's *cantabile* period in the new Duetto

ce — der-ti po - té. Ma in tro - var — ti sven -tu-

- ra - ta, il mio do-no an-cor ri-pren-do, e da

te, da te pre-ten - do l'a-mor pri - mo, la tua

fé, da te, da te pre-ten - do l'a-mor pri-mo, la _____ tua fé.

The accompaniment of the first four verses is for pizzicato strings; at 'ma in trovarti sventurata' they return to a performance with the bow, and provide a fine dominant pedal, sustained also by the winds. Particularly noteworthy in this music is Donizetti's use of appoggiaturas and melodic chromaticism, which cast a melancholy veil over the entire passage. The melodic type represented here appears on other occasions in Donizetti's music; there are examples in both *Lucia di Lammermoor* (Edgardo's 'Sulla tomba che rinserra' in his Duetto with Lucia in the first act) and *Belisario* (the Cavatina of Antonina, 'Sin la tomba è a me negata'), both pieces in G minor and 3/8.

The cadences *a 2* of this Larghetto, on the other hand, cost Donizetti much effort. In the original version we find a repeated eight-measure phrase, with a rather banal continuation. The latter opens *forte*, with a cadence accompanied by the entire orchestra, then closes with an unaccompanied cadence for Anna and Percy. Here the text is so badly distributed as to constitute a serious compositional blunder (see Example 46(a)):

Ex. 46. Cadential alterations in the *cantabile* of the new Duetto

(b)

Recognizing the problem, Donizetti canceled the entire passage, at one point intending to omit even the preceding eight-bar cadence. Fortunately he immediately repented this latter cut, and added the words 'si fa questo'. Then he rewrote the final cadence, as in Example 46(b), conserving the notion of an unaccompanied passage but finding a far superior musical garb in which to clothe it.

There is another interesting point to make about this Larghetto, one that involves the text. Writing Anna's strophe, Donizetti composed music for the first five verses. For the sixth he set not 'Ch'io con te morrei contenta' but rather 'Io lo giuro al Ciel che m'oda'. We do not know the seventh verse of the original poetry because in the autograph Donizetti left it out. The final verse was the same as in the definitive version with one difference: instead of 'Questa sol' Donizetti wrote 'Questo sol'.

The invocation of God in the original sixth verse suggests that the poetry may have continued with words the composer feared might meet objections from the censors. Rather than face such an eventuality, he left the words out. Later, another hand, not that of Donizetti, supplied the missing verse, changed the sixth verse, and substituted the feminine 'Quest*a*' in the last. When we think about the nature of censorship in Italian opera of this period, we have to remember that composers kept these matters in mind even before formally submitting their work for approval.

In the Allegro *tempo di mezzo* that follows immediately, the characters confront one another again. Their dialogue is short, only eighteen bars long, but intense, modulating from G to F major as dominant. After a strong orchestral dominant, the cabaletta begins. We have already seen how difficult the problem of the cabaletta was for Donizetti at the time he composed *Anna Bolena*. In piece after piece we followed his efforts to write a composition that was not an imitation of Rossinian models. From this point of view, the cabaletta of the new Duetto is surely the most successful one in the opera. It opens with what appears to be a standard cabaletta theme, Moderato, written with great care and refinement:

Ex. 47. Cabaletta theme of the new Duetto

Sung first by Percy, the theme opens in B♭ minor, passes suddenly to the relative major, then returns to the tonic. Notice that the phrase is five measures long; its development seems so natural that no other structure is possible. After a move to the dominant, with a similar phrase reduced to four measures, the mode shifts to major, as the dominant continues to be prolonged. The two cadential phrases are joined together melodically so that they form a single gesture, climbing higher and higher to the climactic B♭, over which Donizetti writes the words 'di petto'. This is a most revealing indication of contemporary vocal practice, since the tenor, Rubini, would probably have otherwise sung the note with a head voice rather than a chest voice. We may recall that a year earlier Rossini composed the role of Arnold in *Guillaume Tell* for Adolphe Nourrit, who sang high notes entirely in a head voice.[2] The difference in attitude reveals a fundamental shift in the taste of the two composers, chronologically so close yet so diverse in many aspects of their art.

After Anna repeats the cabaletta theme without essential change, a 'Più Allegro' seems to be preparing a final repetition of the theme, *a 2*, the standard conclusion for such a duet. It is apparent in the autograph that Donizetti did at first end the piece in precisely this fashion (see Example 48 on p. 170). Indeed the cabaletta also closed this way in an earlier version written the previous year, the last section of the Duetto for Imelda and Bonifacio, 'Ah m'odi! . . . (Qual voce!) . . . Imelda! . . .' in

---

[2] See the entry by Philip Robinson on Adolphe Nourrit in *The New Grove Dictionary of Music and Musicians*, ed. Stanley Sadie (London, 1980), Vol. 13, pp. 431–2. The general problem of changes in vocal technique during this period is treated by Celletti, 'Il vocalismo italiano da Rossini a Donizetti', op. cit.

Ex. 48. Preparation for the repeat of the cabaletta theme originally planned in the new Duetto

Imelda de' Lambertazzi (Naples, Teatro San Carlo, 5 September 1830). Its cabaletta theme, both words and music, is essentially identical to that of the new Duetto of *Anna Bolena*.[3] In *Imelda*, after the moment of transition and preparation (which is rewritten for *Anna Bolena*), the original cabaletta theme is repeated intact, sung by Bonifacio with interjections from Imelda. When Donizetti decided to insert this cabaletta into *Anna Bolena*, he originally planned it with the same conventional structure.

It is not certain how far he actually proceeded with this plan: all that remains of the repeat are three measures on f. 223v (the final three measures in Example 48). But moved undoubtedly by his urge to intensify the dramatic situation rather than to submit willingly to abstract formal requirements, Donizetti crossed out the melody of the reprise and began entering new material directly over it. When the resulting score became too confused, he drew a grid of lines over the three measures and began again, adding a final gathering of two bifolios (ff. 224–7), on which he wrote the new conclusion of the Duetto. Here is the principal part of this conclusion:

Ex. 49. The 'cabaletta of the cabaletta' in the new Duetto

---

[3] Although a complete piano-vocal score of the opera does not appear to have been published, Ricordi did bring out this 'Scena e Duetto' from *Imelda de' Lambertazzi*, with the plate number 5132, 14 pp., at approximately the same time as the piano-vocal score of *Anna Bolena*, early in 1831.

the music concludes with
a group of cadences.

Cadences *a 2* follow, with the structure: $2 \times 4 + 2 \times 2 + (2 \times 1 + 2)$. They lead into a short orchestral transition and directly back to the original Finale $1°$.

This conclusion is electrifying. The shift in tempo to 'Più mosso', the canonic interplay of the voices, the 'crescendo di forza e di tempo', the dramatic moment in which Anna 'abandons herself in the arms of Percy', the repeat of the initial canon with the vocal parts reversed, all combine to render the final moments of the new Duetto for Anna and Percy one of the strongest dramatic inspirations in Donizetti's operas. And it may be noted that, in structural terms, one could well define this passage as the 'cabaletta of the cabaletta'. Such an alternative to the regular cabaletta structure depends in part on the element of surprise. Because our expectations are deceived, we do not hear the conclusion as yet another block of music tacked on at the end. This type of ending, therefore, cannot be considered a truly viable alternative to the cabaletta structure, though there are fine examples in some other operas: Rossini's Terzetto in *La donna del lago* (1819) or the Duetto for Parisina and Ugo, 'Dillo . . . tel chieggo in merito', in Donizetti's own *Parisina* (1833). Were the force of our expectations weakened, however, we would be left with a simple additive musical structure, with the problems of comprehensibility and integrity that such forms inevitably raise.

Given our present knowledge of Donizetti's operas, it is impossible to follow with precision the further history of 'Sì, son io che a te ritorno'. That it was occasionally sung in later years is revealed, as we shall see, in the correspondence of Alessandro Lanari, impresario of the Teatro della Pergola in Florence. Though it might be considered strange that such a beautiful piece did not become an integral part of the opera, the reason seems self-evident and has been attested to as well by modern singers: the Duetto, dramatic and fully developed, is extremely difficult for the singers and precedes a long and demanding Finale, especially for Anna. Donizetti may well have become aware of this problem and hence, despite the composition's inherent qualities, it did not normally figure in *Anna Bolena*.

Yet it would be wrong to say that it disappeared completely. Sections of this Duetto reappeared in the opera Donizetti wrote for the Théâtre-Italien of Paris, at the invitation of Rossini, *Marino Faliero* (Paris, 12 March 1835).[4] And that at least the

---

[4] The textual history of *Marin Faliero* is quite complex. I have tried to sketch some of the issues involved in my study 'Music at the Théâtre-Italien', to be published in the proceedings of the conference held at Smith College in the Spring of 1982, *Music in Paris in the Eighteen-Thirties*. The Duetto borrowed in part from the new Duetto of *Anna Bolena* was not contained in the opera Donizetti brought with him to Paris; rather it was added as one of the revisions the composer made in Paris.

Florentine public recognized the derivation is apparent from the letter of Lanari to Donizetti dated 19 May 1836.[5] After reporting that 'Your *Marin Faliero* was judged to be your masterpiece', Lanari, reviewing each piece in order, wrote: 'The Duetto between Elena and Ferdinando was sufficiently applauded, but the public recognized in its two first sections those of the Duetto from *Anna Bolena*, performed so many times.' Thus, in Florence at least, our Duetto was well known.

The first section of the Duetto from *Marino Faliero*, 'Tu non sai, la nave è presta', is in fact essentially the same as the first section of the Duetto from *Anna Bolena*. Of the second section, only the first phrase is borrowed from the earlier piece: the rest is newly composed. The strangest aspect of this case of self-borrowing is that the version of the opening section taken over in *Marino Faliero* is unquestionably weaker than the original. In *Anna Bolena* we are struck by the diversity of rhythmic designs, by the melodic and dramatic variation that defines the characters precisely: Percy young and naïve, Anna already hard and prematurely aged by her unhappiness. In *Marino Faliero*, on the other hand, the characteristic differences disappear. The dotted rhythm used with such force by Donizetti to express the words 'Dell'amor l'età trascorse' in Anna's strophe appears indifferently four times in Fernando's stanza, and then again in Elena's. The delicate original chromaticism is replaced by trivial figuration, as in the very first phrase:

Ex. 50. The opening of the Duetto for Elena and Fernando in *Marino Faliero*

[5] Given in Zavadini, *Donizetti*, pp. 867–8.

Elena's strophe, apart from some very minor changes, is no longer a melodic variant of Fernando's. Passion has disappeared, individuality has disappeared. It is difficult to imagine how Donizetti could have changed the piece in this way, unless perhaps he was recalling it from memory. But instead of the correct tone, he found only a pallid reflection of his earlier inspiration. At least part of the fault may lie with the characters in *Marino Faliero*: neither Elena nor Fernando is drawn in the libretto with the skill that Romani lavished on Anna and Percy. Indeed it is hard to sympathize with these characters, wife and nephew of the Doge, who have already committed adultery. In any case, it remains a fascinating example: from the later version one could have little sense of the extraordinary qualities of the original.

The revised Duetto for Anna and Percy is a superb composition. What makes study of its autograph fascinating, though, is that many of the elements that give the piece its character are the product of compositional revisions still visible in Donizetti's manuscript. They reflect precisely the host of revisions we have been following throughout the opera: changes in melodic detail, revisions in transitional sections, omissions of formal orchestral introductions in favor of more dramatically pointed confrontations and greater musical continuity, the search for alternatives to the traditional cabaletta design. Even beyond these specific factors, though, there remains the image of Donizetti, the composer, listening, judging, reacting, driven by his own internal needs, by the forces of history and convention, by his empathy with his characters. However much we can recreate those processes through which the composer arrived at his finished opera, we must always react with wonder at the miracle by which these varied and even contradictory elements can come together to produce a masterpiece.

XII. Making Form Responsive to Content

ONLY once in the autograph of *Anna Bolena*, in the Introduzione (No. 1), can we watch Donizetti struggling with the basic shape of an entire composition. But within the boundaries of convention that circumscribe his art at this moment in his life, his solutions here are notably successful. The 'Introduzione' encompasses all the music from the opening chorus to Anna's cabaletta, 'Non v'ha sguardo cui sia dato'. That Donizetti conceived it in this way is shown by the label he affixed to the recitative preceding the Duetto for Giovanna and Enrico, 'Dopo l'Introduzione'. What distinguishes this composition is the way Donizetti introduces in turn Giovanna, Smeton, and Anna, characterizing each musically with great skill.

Giovanna's entrance in particular is unusual in its fusion of dramatic recitation ('Ella di me, sollecita') and lyrical expression ('Innanzi alla mia vittima'). It is noteworthy that when he first wrote out the accompaniment to the more lyrical part of Giovanna's entrance, Donizetti continued precisely the same tremolo accompanimental figuration he had adopted earlier. Example 51 shows a few measures of this accompaniment, together with the later revision (see p. 178).

The alteration in the string figure is obviously central to the way we perceive the lyrical qualities of Giovanna's melody, and hence to the sympathy we feel for her character at the very start of the opera. Smeton's song, 'Deh! non voler costringere', on the other hand, is a strophic Romance, of the kind that dominated French grand and not so grand opera during the first part of the century, but which played a smaller role in the Rossinian scheme of things.[1] This Romance defines Smeton much in the

---

[1] There are Rossini examples, though, such as 'Se il mio nome saper voi bramate' from *Il barbiere di Siviglia*, while 'Sombre fôret' from *Guillaume Tell* may well be the finest Romance in all French opera.

Ex. 51. Revisions in the accompaniment of Giovanna's entrance in the Introduzione (No. 1)

way 'Voi che sapete' had defined Cherubino in *Le nozze di Figaro* and as Oscar's little songs would define him in *Un ballo in maschera* some thirty years later. Anna is presented in more conventional terms, but they unmistakably cast her as prima donna *extraordinaire*.

The shape of the Introduzione, diagrammed in Table XIII, formed only gradually in the minds of Donizetti and, presumably, Romani:

Table XIII. General structure of the Introduzione (No. 1)

Section	Key	Paper
(A) Coro d'Introduzione	E♭ major	1
(B) [Sortita] Giovanna Seymour	→A♭ major	2
(C) Scena		2
(D) [Romance] Smeton	E♭ major	1
(E) *Cantabile* Anna, 'Come, innocente giovane'	G major	2
(F) *Tempo di mezzo*	V of E♭ major	2
(G) Cabaletta Anna, 'Non v'ha sguardo'	E♭ major	2

With the documents now available, it is impossible to trace all the steps, but the autograph gives certain information. Notice that two different paper types are used in this Introduzione. Type 2 is a twenty-stave paper that serves as the basic paper for most of the autograph of *Anna Bolena*. Type 1 is a sixteen-stave paper, found only at the beginning of the opera: it is the paper for the main body of the Sinfonia (the opening section is on type 2) and for a single page in Percy's Cavatina (f. 99, the page that may reflect an earlier version of the cabaletta, as discussed in Chapter VIII).[2] Type 2 reappears in the Duetto for Giovanna and Enrico (No. 2); later, from the Finale 1° through the entire second act, this paper is found almost exclusively. These factors certainly suggest that the Coro d'Introduzione (A) and the Romance of Smeton (D) are part of an earlier version of this Introduzione whose entire physiognomy we cannot reconstruct but which we can know in part. Though the Romance of Smeton begins on f. 35v of the autograph, f. 35r is *not* the end of the recitative (C), which we believe to have been added later. Rather, it is the close of a *cantabile* movement in 6/8, presumably for Anna (see Example 52 on p. 180). Preserved only with the vocal line, bass, and occasional violin material, this *cantabile* was surely omitted early in the opera's history. Not much text is present, but 'da me non ti scostar' is presumably an invocation, aside, by Anna to Enrico. This movement was followed by Smeton's Romance, but the manuscript on paper type 1 continues only through its second full

---

[2] Overtures were usually composed last, so that we cannot draw many conclusions from the places paper type 1 is found. The nature of the revisions present on this paper in the Introduzione, however, favours our hypothesis.

Ex. 52. Autograph remnant of an earlier *cantabile* for Anna, before Smeton's Romance in the Introduzione (No. 1)

stanza, stopping before the E♭ major resolution at 'quel primo a-[mor]'.[3] The last measures of Smeton's Romance are on the following folio, paper type 2.

Now, if in an original version Anna had sung a *cantabile* movement before Smeton's Romance, she would not have sung another after it. But instead of this earlier *cantabile* referring to Percy, it spoke of her sorrow at Enrico's coldness. In the final version of the Introduzione, we have little reason to think Anna has any feelings at all for Enrico. Perhaps in the original version she showed some in this *cantabile*. Smeton's Romance may then have intervened, calling Percy, her 'primo amor', to mind, and this may in turn have motivated a cabaletta different from the one we know today.

[3] The original printed libretto makes no mention of Smeton beginning a third stanza with the text 'Quel primo amor che . . . ,' and it was presumably Donizetti's idea, an excellent one.

Though this is all largely hypothetical, to be sure, it is necessary to account for those measures preceding Smeton's Romance and for the diverse paper types. There is another indication that the Introduzione was put together in layers. The recitative scene (C) has written across it, in Donizetti's hand, 'Dopo il primo Coro / Atto I'. If Giovanna's entrance had been present from the start, even from the start of the layer associated with paper type 2, for section (C) is also on paper type 2, the nomenclature would be absolutely incorrect. Giovanna's entrance may be structurally unusual and one may have to search for a way to refer to it, but there is certainly no justification for calling it 'primo coro'. Furthermore, Giovanna's entrance is found on ff. 28 to 30r, with f. 30v blank. Had Donizetti been composing continuously, he would not have left this page blank.[4] On f. 28r someone else wrote 'Dopo il P$\overline{\text{mo}}$. Coro. Atto P.°', precisely what Donizetti himself had originally signaled at the start of section (C), which begins on f. 31r.

It seems quite clear, in short, that the definitive structure of the Introduzione was not the original plan of the piece. The original probably lacked Giovanna's entrance and used Smeton's Romance as a *tempo di mezzo* between a *cantabile* and a cabaletta for Anna. In revising this Introduzione, Donizetti comes to grips with the formal conventions of his time and succeeds, in a surprisingly simple fashion, in adapting them to the needs of introducing and differentiating characters at the very opening of the opera. He pushes beyond the limits of the inherited structural conventions, while not abandoning them.

---

[4] Compare the situation mentioned above, where the close of Anna's original *cantabile* is on f. 35r and Smeton's Romance begins immediately on the verso.

Conclusion

It has been the purpose of this study to examine critically those passages in the autograph score of Donizetti's *Anna Bolena* that show the composer in the act of making compositional decisions: decisions affecting phrases, periods, entire sections, and entire compositions; decisions concerning details, proportion, continuity, and large-scale structure. In each case I have attempted to define why Donizetti moved instinctively to certain musical devices, why he backed away from them, and why his final solutions may have seemed more appealing. I do not think these ultimate solutions are always best, but Donizetti was not sitting as an objective judge. He was a composer living at a certain time with a very strong inherited tradition, a tradition he accepted but whose constraints seemed at times too restrictive.

In practically every case examined, Donizetti's instinctive formal decisions return to Rossinian models, the common denominators of Rossini's style. By 'formal decisions' I mean not only the overall structure of compositions, but the way that individual phrases and periods attain coherence and balance. His alterations express impatience with these models: he upsets the proportions, runs sections into one another, and avoids the larger structural imperatives. These alterations do not always get at the heart of the matter. Lopping stones from a Greek temple does not change its style, but the eye, once accustomed to the resulting asymmetries, may be better able to work out a new style. In this sense, the changes in the autograph of *Anna Bolena* are significant. Though the opera is conceived and executed under the formative influence of Rossini, we can watch Donizetti tentatively pressing against his boundaries. In some cases he is extremely successful in personalizing the structures within which he is working; in others he lashes out at them without escaping their hold.

Surely this process of artistic maturation does not begin with

Anna Bolena, nor does it end at any specific moment in the composer's life. Careful study of the earlier operas reveals not only much beautiful music but also much music in which this same process of artistic maturation can be followed. Only when we truly come to know these works will we be able to assess precisely the position of *Anna Bolena* within this longer process.

Until then, at least, I hope to have shown that it is possible to catch a glimpse of the composer in 1830, to share his musical concerns, and to understand better how *Anna Bolena* reflects his growing maturity.